# Power, Participation, and Policy

# Power, Participation, and Policy

## *The "Emancipatory" Evolution of the "Elite-Controlled" Policy Process*

Katsuhiko Masaki

LEXINGTON BOOKS

A division of
ROWMAN & LITTLEFIELD PUBLISHERS, INC.
*Lanham • Boulder • New York • Toronto • Plymouth, UK*

LEXINGTON BOOKS

A division of Rowman & Littlefield Publishers, Inc.
A wholly owned subsidiary of The Rowman & Littlefield Publishing Group, Inc.
4501 Forbes Boulevard, Suite 200
Lanham, MD 20706

Estover Road
Plymouth PL6 7PY
United Kingdom

British Library Cataloguing in Publication Information Available

**Library of Congress Cataloging-in-Publication Data**

Masaki, Katsuhiko, 1964-
  Power, participation, and policy : the "emancipatory" evolution of the "elite-controlled"
policy process / Katsuhiko Masaki.
    p. cm.
  Includes bibliographical references and index.
  ISBN-13: 978-0-7391-1177-2 (cloth : alk. paper)
  ISBN-10: 0-7391-1177-9 (cloth : alk. paper)
  1. Political participation--Nepal--Case studies. 2. Flood control--Nepal--Citizen
participation--Case studies. I. Title.
  JQ628.A1M37 2007
  363.34'936095496--dc22                                            2006028381

Printed in the United States of America

∞™ The paper used in this publication meets the minimum requirements of American
National Standard for Information Sciences—Permanence of Paper for Printed Library
Materials, ANSI/NISO Z39.48–1992.

To

My Parents

Yusaku and Yuriko Masaki

# Contents

# Table and Figures

# Maps and Photos

# Abbreviations

| | |
|---|---|
| CPN-UML | Communist Party of Nepal-Unified Marxist-Leninist |
| D&P | Decentralization and Participation |
| DDC | District Development Committee |
| DIO | District Irrigation Office |
| DOI | Department of Irrigation |
| NC | Nepali Congress Party |
| RPP | *Rashtriya Prajantra* (National Democratic) Party |
| VDC | Village Development Committee |

Glossary of Nepali Terms

| | |
|---|---|
| *badghar* | leader of *Tharus'* communal labor system |
| *begaari* | unpaid labor |
| *dalit* | Hindu low caste group/untouchables |
| *Dheshi* | Hindu caste groups originating from the Indo-Gangetic Plain |
| *kamaiya* | bonded laborer |
| *khel* | *Tharus'* communal labor system |
| *Pahadi* | hill people |
| *Panchayat* | partyless monarchy preceding the multi-party system in the 1990s |
| *sukumbasi* | landless settler |
| *Tarai* | narrow lowland strip stretching along the border with India |
| *Tharu* | ethnic group from Western *Tarai* |

# Acknowledgments

This book is based on my Ph.D. dissertation ("The Politics of the Policy Process: 'Participatory' River Control n Nepal") submitted to the Institute of Development Studies (IDS) at the University of Sussex (2003). I owe my deepest gratitude to Jude Howell who was a conscientious and excellent supervisor. Without her attentive guidance, my research journey would not have been such a heartening experience. She has also played an invaluable part in the production of this book, with her unwavering intellectual and moral support.

Special thanks are also due to people in Majuwa and Pyauli, the names of whom I cannot disclose to maintain confidentiality. Given that the emphasis of my research was to delve into micro-level power struggles, I may unwittingly have asked intrusive and offensive questions at times. Still, the residents were generous enough to speak their minds about contentious issues. Not only in the two areas, but both in the Bardiya district and in Kathmandu, I interviewed a number of politicians, government officials, and others, who not only gave me their time but were open with me. Some of them asked for anonymity and it is for this reason that I refrain from mentioning their names. It was through my interactions with various policy "elites" and their "clients" that theoretical literature has come to register in my mind with startling reality. I owe my grip on abstruse theories to those who candidly imparted their life worlds.

My friends in Nepal were also an important source of inspiration. Kiran Amatya, a seasoned community organizer, shared with me an insider's insight into the symbolic and manipulative side of "participatory" development. Lok Prasad Bhattarai, a sociologist of the Department of Irrigation (DOI), took a special interest in my research, and often helped me out of an impasse by providing fresh syntheses of my fieldwork results. I also appreciated my interactions with Krishna Hachhethu, a noted political commentator, who constantly reminded me of a possible hidden agenda lurking behind flowery rhetoric of Nepalese politicians. Mukti N. Manandhar from the DOI helped me, beyond the call of his duties, to make contact with senior politicians and bureaucrats, with whom I would otherwise have faced difficulties in making appointments. Masamine Jimba and Izumi Murakami, from the JICA-supported School and Community Health Project, generously offered me office space in Kathmandu, while I was preparing for my fieldwork in the Bardiya district. During my field work in Bardiya, Fumi and Cliff Meyers, and Jamuna Kayastha helped me cope with the competing demands to attend to field research and my personal problem. I must also express my sincere thanks to Rabindra Bahadur Shah, a socially-minded "jamindar", who taught me the everyday challenges that a social worker like him faces in Western Nepal. My research assistants from Bardiya, Lok Bahadur Tharu, and Vikram Sharma, not only provided excellent logistical support, but also shared with me their life experiences in the district, which provided invaluable clues in gaining a deeper understanding of the district.

At the Institute of Development Studies (IDS), I am thankful to James Manor, who initially invited me to the IDS. Other fellows and staff members at the IDS provided crucial inputs in relating my fieldwork results to contemporary thinking, including John Gaventa, Anu Joshi, James Keeley, Sarah Lister, Lyra Mehta, Garett Pratt, and Ian Scoones. I must also acknowledge the friendships and encouragement extended by my D.Phil. colleagues, especially those of Paul Howe, Luka Biong Deng, Colette Solomon, and Fang Jing, who provided useful sounding boards for my emerging ideas.

The fieldwork for this book coincided with a time when my personal life was in turmoil. In this respect, I would also like to express sincere gratitude to Sue Ong, and Marc Fiedrich, who showed very special concern about my well-being, and cheered me up by keeping on assuring me that there is always "light at the end of the tunnel" and that "every cloud has a silver lining."

This research was made possible by funding from the Foundation for Advanced Studies on International Development (FASID). I was hosted in Nepal by the Centre for Nepal and Asian Studies (CNAS), at Tribhuban University in Kathmandu. I am extremely grateful for their generous assistance.

I would also like to express my deep appreciation for Rebekka Brooks, my original acquisitions editor at Lexington, for encouraging me to write this book, and Molly Ahearn and Joseph C. Parry for seeing it through the production, and for their meticulous attention to the text which greatly improved the final product.

Finally but not least, my greatest debt of gratitude is to Ayako and Sho. Their presence, support and love have been an invaluable source of strength.

# Chapter 1
# Introduction

"If I were a strong politician, I would tell the DG [Director General of the Department of Irrigation] to construct irrigation canals in my village. Even if the DG advised there was no water there, the Minister would say 'It does not matter. I want my irrigation system there.' . . . So, the best way is to establish clear-cut procedures. Suppose your son fails an exam, do you or does the Minister go to the teacher and say 'You have to pass my son'? No. Although the Minister is very powerful, there is a standard, objective process of judging the performance of your son. So, there is no point in putting political pressure on the teacher."[1]

"We have a good policy on paper. We have yet to implement it fully. It is an open secret that politicians are indulging in corruption. And our [government bureaucrats'] mentality has not changed yet. Still many of us would like to go the way we did before [disregarding 'people's participation']. Our major problem is not policy, but implementation. Our policy says 'We should involve people', but we are not doing it to the extent we should. It is improving slowly, but not fast enough."[2]

These statements reflect several assumptions, widely held among policy analysts, about popular participation in the policy process. The first narrator from a donor agency in Nepal refers to the significance of prescribing objective criteria to discipline political leaders, who are liable to impose programs incongruent with local circumstances. The second speaker, a government official in Nepal, describes how the country's policy on "farmer-managed" irrigation is plagued by the lack of commitment on the part of those in official positions, on the assumption that ordinary people are submissive to decisions taken at higher levels. These presumptions together conjure up orderly world-views, namely that high-level decision-makers are disinclined to relinquish their influence in policymaking, that the general public exercises little leverage over the unfolding of the policy process, and that it is imperative to take countermeasures against elite-controlled processes of policymaking.

Accordingly, one lasting theme in policy studies is to explore how to ameliorate the manipulative nature of policymaking that overrules or redirects the desires and aspirations of ordinary people. In this respect, people's participation in the policy process has emerged as a converging concern of several themes in development studies. There are increasing calls from researchers on participatory development to move away from the "user group" approach that reduces people to "consumers" of public services, towards broader people's involvement in

1

policymaking.[3] The "good governance" agenda has extended beyond the conventional focus on the delivery of government services, to highlight the importance of popular engagement in major policy decisions, in order to hold government institutions accountable to the public.[4] In parallel with the rise of the "rights-based approach to development" is the re-conceptualization of "citizenship" to encompass citizens' rights to participate in the policy process that affects their lives.[5]

"Citizen" participation, or broader people's engagement with policymaking, is often called for by proponents of popular participation.[6] Advocates of "citizen" participation make crucial contributions by attracting our attention to a broader terrain of participation that has been, and can potentially be promoted, than is the conventional approach, which equates participation narrowly with a project-based intervention in a bounded locality. As pointed out by Hickey and Mohan, it is imperative not to regard "the spread of PRA" as the definitive modality of participation, but to pay attention to previous debates around various alternative approaches to development.[7] Cornwall similarly proposes to link popular participation with wider reforms of promoting "civil society", "decentralized local governance", and "rights and social justice."[8] In their views, in order to rejuvenate the notion of participation, it is imperative to move away from the conventional "user group" approach that merely aims at getting people to take on the execution of external projects, towards "citizen" participation catalyzing transformation in the overarching policy process.

This study calls into question such a binary view of "user group" (project-based) and "citizen" (broader forms of) participation, by drawing on a post-structuralist notion of power which elucidates the co-eixstence of oppressive and emancipative elements in social control. Underlying the notion of "citizen" participation is a simplistic, dichotomized view, ascribing power to dominant actors who exercise it over others. According to proponents of "citizen" participation, in line with the pervasive assumptions about policy, referred to at the beginning this chapter, elites are liable to impose policies incompatible with local aspirations, and ordinary people are subjugated to such exclusionary dynamics. It is therefore imperative, so the story goes, to launch a distinct set of initiatives to turn the general public into fully-fledged "citizens" capable of engaging in policymaking. But in the light of the subtle nature of power dynamics in which opression and emancipation are interwoven, is it true that people would not be "citizens" under the "user group" approach? If both despotic and transformative forces are entangled in social interactions, does not this mean that policy "clients", even when they are relegated to the status of "users" of a discrete project, can potentially exercise leverage and participate in shaping the overall unfolding of policymaking?

One prominent post-structuralist view of power is the conception of "disciplinary" power proposed by Foucault, who reveals how power subjugates people into constrained positions by exerting pressure on them to conform to societal standards.[9] The Foucauldian notion of power refutes the above-mentioned accepted presumptions about policy. First, at the nucleus of the Foucauldian conception is the contingent and inconsistent nature of individuals' subject positions, which are constantly being adjusted and modified through daily social interactions. Second, power not only constrains individuals' thoughts and actions, but also serves as a medium through which different social actors renegotiate their interpretations of reality. As pointed out by Foucault, therefore, where and when power is in operation, a "whole field of responses, reactions, results, and possible inventions may open up."[10] These assertions by Foucault lead us to a different understanding of policymaking, namely that the policy process is not entirely controlled by politicians, bureaucrats, and specialists, the unfolding of which is contingent on how various policy stakeholders interact with one another in particular circumstances.

This Foucauldian productive conception of power manifests itself as a strand of policy studies in what Shore and Wright term the "anthropology of policy."[11] In development studies, the "anthropology of policy" is adopted by "post-development" thinkers,[12] who regard development policy as a discourse that fabricates reality in such a manner as to "discipline" the general public to embrace the expansion of state intervention, or the "governmentalization" of society. In line with the above-mentioned two major features of the Foucauldian notion of power, at the same time, the "anthropology of policy" considers that policy elites do not necessarily work to entrench their clients' dependence, and that the general public are not passively subjugated to the exclusionary nature of policymaking. A policy intervention, in however top-down a manner it may be made, offers an arena where beneficiaries reacts to, and resists decisions taken at higher levels, and decision-makers ceaselessly reflect on, and re-adjust the policy direction. Proponents of the "anthropology of policy" therefore point out emancipative potentials inherent in the daily flow of policymaking, unlike those others who seek to reconstitute policies with a view to fully unleashing their progressive potentials. They accordingly seek a "utopian possibility of re-conceiving and reconstructing the world from the perspective of, and along with, those subaltern groups that continue to enact a *cultural politics* of difference."[13]

## Hypotheses, Background, and Research Questions

This study seeks an unorthodox approach to popular participation in the policy process, by viewing policymaking as a complex web of "cultural politics" taking place in a myriad of places. If policymaking is implicated in a dynamic meshing of varying meanings attached by different stakeholders, it is crucial to gain a more profound understanding of the exclusionary mechanisms facing policy "clients", than is present in the orderly world-views underlying the notion of "citizen" participation. If the "disciplinary" power entailed in policymaking not only constrains but also facilitates public engagement in decision-making, ordinary people are not entirely submissive to its exclusionary forces, while high-level policy-makers do not always seek to defend their influence in policymaking. The policy process should instead continue to be reshaped and remade by ongoing power renegotiations, through which the general public can exercise influence over the way policy is initiated and carried out. Even under the "user group" approach which reduces people to the role of "consumers" of external interventions, policy "clients" potentially exert some leverage over higher-level decision-making.

The key hypothesis of this research is therefore that a potential for the emancipation of marginal people, ostensibly "barred" from policymaking, would be immanent in the daily flow of the policy process, in contrast to existing studies that assume that a separate process of countermeasures should be launched. I also examine two sub-hypotheses that qualify the above main hypothesis. Given the multi-faceted, fluid nature of individuals' identities, the policy process should take on a contingent and transient nature, rather than a regular and predictable form that can be deduced abstractly from the structural locations of the respective policy actors. First, therefore, it is surmised that the interface of the top-down policy directive and local beneficiaries cannot be reduced to a simple process of assimilation and resistance. Second, the binary logic that categorizes various stakeholders into "goodies" and "baddies", as is often implied in studies on popular participation in policymaking, is too simplistic to capture the actual un-folding of policymaking.

By examining these hypotheses, this study primarily aims at criticizing the prevailing orderly views that policy "elites" are liable to impose policies incon-gruent with local situations, and that countermeasures should be taken to ame-liorate the distorted process of policymaking and to promote public participation in policy decisions. For this purpose, this book draws on the case of a policy di-rective that was issued by the Government of Nepal in 1992 to realign local river control projects with the emergent decentralization and participation (D&P) strategy. Nepal is divided into three physiographic zones, the mountains, the hills, and the Tarai Plain (see Map 1.1). In Nepal, monsoon clouds bring torrential rainfall to the southern slopes of the mountains and the hills, which causes flooding in the southern Tarai. To cope with flooding in the Tarai, the Govern-ment of Nepal has been implementing national and district-level projects. Na-tional-level river control projects normally take the form of continuous dikes along stretches of rivers. District-level river control projects, the subject of this study, usually target a limited number of settlements, through the construction of small-scale structures, such as groins or spurs to reduce water velocities and di-vert flows away from riverbanks (see cover photo), and retaining walls to protect riverbanks.

Before the policy directive was issued in 1992, the central government op-erated its district-level river control portfolio through its deconcentrated offices. The 1992 river control policy appointed the chair of the district governing body,

**Map 1.1. Nepal**

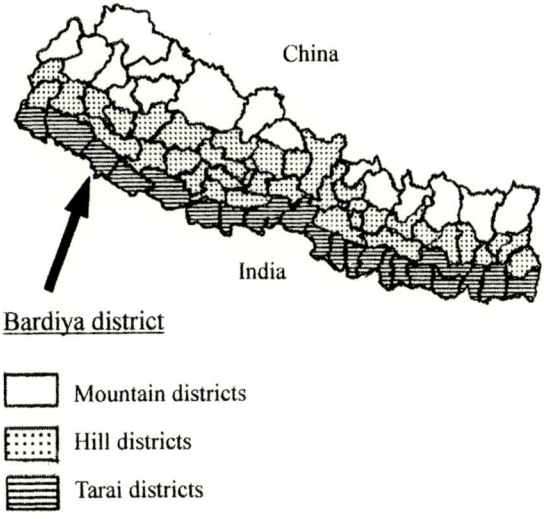

Bardiya district

[ ] Mountain districts

[:::::] Hill districts

[≡≡≡] Tarai districts

to head a district-level committee for flood mitigation. Though the center was to retain control over the allocation of resources to different districts, where and how to use the allotment was to be left to the district-level committee. The policy also stipulated that district-level river control projects should be undertaken with the mobilization of labor contributions from the beneficiaries. However, in the latter half of the 1990s, the "participatory" policy came to be ignored by the center, which instead imposed its own decision about a project plan. In the late 1990s, the central government often gave instructions to use contractors in lieu of labor contributions, and specified the locations and types of river control structures, instead of sending a blanket authorization for each district to use the budget at will.

A conventional approach to analyzing the decline of the policy would be to examine the factors that caused "power-holders", or central-level political leaders, to belittle the policy. Underlying this type of "descending" analysis[14] is a simplistic, dichotomous conception that power is exercised by dominant players over oppressed actors. However, according to the main hypotheses of this research, it is assumed that policy "clients", or flood victims in this case, were not simply subsumed under decisions taken by high-level policymakers, but instead exercised some leverage over the unfolding of the policy. Accordingly, this book does not focus merely on "power-holders", nor deliberate over how their strategies caused the policy to fall into decline. On the other hand, in view of the intricate nature of power that cannot be attributed only to "dominant" actors, this study undertakes an "ascending" analysis,[15] namely examining the downward trend of the policy as a meshing of "cultural politics" experienced by different actors, ranging from villagers to central politicians.

As would be presupposed by the "anthropology of policy", the "participatory" river control policy regulated and disciplined the way in which people viewed flood mitigation, in such a manner as to prompt them to equate it with a top-down process of distributing pre-packaged projects. In this regard, the D&P rhetoric played a crucial role by invoking an image of concern about the well-being of the flood-affected populace, while masking other options that would better unleash people's own initiatives for flood mitigation. In line with the notion of "cultural politics", at the same time, I presume that the constrictive notion of river control was ceaselessly reflected upon by various stakeholders at different levels. It is thus surmised that an emancipative potential for redressing the manipulative nature of policymaking was immanent in the daily execution of the "participatory" river control policy. To make sense of the rise and fall of the policy, therefore, the book avoids analyzing the policy process abstractly from reified representations of social actors. It instead makes empirical inquiries into the dynamic interfaces among various stakeholders, including grassroots beneficiaries and central decision-makers, the results of which were presumably uncertain and even at times "unexpected."[16] In doing so, this research addresses three key questions, described in the following section.

## Key Research Questions

The study first inquires into the discursive practices of national policymakers that brought about the "participatory" river control policy in 1992. Despite the rise of the decentralization and participation (D&P) agenda in the 1990s, a "story-line" that treated development as exogenous to "remote" villages continued to set the overall tone of the conduct of development in Nepal. Against this background, I examine how the D&P rhetoric, overridden by the top-down "story-line", was incorporated into the river control policy, in such a way as to relegate it to the delivery of pre-packaged handouts, rather than unleashing people's own initiatives. At the same time, in view of the provisional and incoherent nature of policymakers' identities, this research delves into how the ostensibly immutable "story-line" was subject to constant reflections among policymakers with unstable subject positions.

The study then addresses the question as to how various social actors and the national directive of "participatory" river control intersected in a particular local context. It does this by focusing on two case studies of river control projects in the Bardiya district in Western Nepal. Bardiya is situated in the southern Tarai belt, and is the third district from the western border with India (see Map 1.1). In

Bardiya, the D&P rhetoric was appropriated by local elites as a euphemism for the historical corvée practice that fell inequitably on marginal segments of the local populace. In the light of the Foucauldian notion of productive power facilitating renegotiations of social norms, I avoid simplistically presuming the unqualified predominance of the policy directive over the general public, and instead examine how different local actors reacted to, and resisted the corvée-based modality of "participatory" river control.

Based on the results of the above inquiries, this study finally investigates how one can make sense of the unfolding of the "participatory" river control policy, as a meshing of micro-power contestation among various stakeholders, ranging from villagers to central politicians, who attached different meanings to the policy. If individuals' thoughts and actions are not always driven by clear-cut motives, the interface of the policy and local life worlds cannot be reduced to a binary process of assimilation and resistance to the top-down policy directive. National-level policymakers were not a monolithic entity seeking to manipulate the D&P rhetoric to inhibit public participation in flood control, while local elites did not always seek to suppress marginal groups by assimilating the policy to the local corvée tradition. I therefore pay special attention to how the multi-faceted and fluid nature of the subject positions of national, district, and village-level actors opened avenues for the general public to participate in shaping the evolution of the overarching policy process.

## Field Research and "Reflexive" Methodology

Before concluding this chapter with an outline of the overall structure of the book, I hope to describe the significance of the "reflexive" fieldwork methodology, with reference to my fieldwork in the Bardiya district. I undertook my field research in Nepal from June 2000 to May 2001. My research in Bardiya extended from October 2000 to May 2000, which had been preceded by my national-level inquiries from June 2000 to October 2000. I also revisited Nepal in July 2002, to conduct additional interviews with central policymakers. In Bardiya, in order to trace the evolution of the policy process over the years, I selected two neighboring villages (of which real names are withheld to maintain confidentiality), Majuwa and Pyauli, where river control projects had been conducted consecutively. The government had implemented river control projects in 1993, 1994, and 1996 in Majuwa, and in 1999, 2000, and 2001 in Pyauli.

One strategy I adopted, to maneuver myself into a position in which I could interact with marginal groups, was to use questionnaire surveys,[17] and the choice of this method did not seem to cause much suspicion among "gatekeepers", that is elected and other village leaders. The questionnaire surveys were accompanied by informal conversations that allowed me to identify and build rapport with a number of villagers who were pleased to divulge stories of their plight. I subsequently revisited them for further interactions. To acquire an in-depth under-standing of village power dynamics, I "hung out" with them as much as possible, engaging in casual interactions and observations. Such "participant observation" enabled me to obtain information that I could not solicit through formalistic re-search, and helped me to resist the temptation of jumping to conclusions.

These strategies as a way of gradually grasping the "ground reality", however, are rooted in "naturalism", which advocates examining the social world in its "natural" state. "Naturalistic" inquiry aims at enabling an outsider to gain the perspective of the "natives" through close and long-term contact, which serves both to reduce "reactivity", that is the influence of the observer on the people studied, and to elucidate the meanings guiding their behavior. At the same time, "naturalism" is founded on "naive realism", namely that there is a "reality" in-dependent of the observer, and that the task of the researcher is to represent that "factual" phenomenon accurately.[18]

## "Reflexive" Fieldwork Methodology

While capitalizing on the strength of "naturalism" to capture societal dynamics with "fidelity", I have pursued a more subtle form of realism by constantly re-minding myself that all research is assumption-laden, and that its results are de-rived through selective observation and interpretation. "Naturalistic" inquiry is "reflective"[19] in its endeavor to monitor and reduce the influence of the re-searcher's presence on the interviewees. On the other hand, I have sought "full and uncompromising self-reference",[20] not only by accepting that no research is immune to the respondents' "reactivity", but also by recognizing roles played by the interpreter in the production of study outcomes. To implement such "total reflexivity", I adopted "quadri-hermeneutics" (see Table 1.1), which explore multiple descriptions of social phenomena, drawing on various theoretical posi-tions, namely naturalism, interpretivism, critical theory, and post-modernism.[21]

### Table 1.1. Levels of Interpretation in "Quadri-hermeneutics"

| | |
|---|---|
| 1. | Interaction with empirical materials |
| 2. | Interpretation |
| 3. | Critical interpretation |
| 4. | Reflection on text production and language use |

Source: Alvesson, Mats, and Kaj Skoldberg. *Reflexive Methodology:
New Vistas for Qualitative Research*, London: Sage, 2000, 250.

To illustrate this "reflexive" methodology, the rest of this section is devoted to describing how I pursued multi-level reflections in my village-level fieldwork.

As mentioned below in chapters 5 and 6, in the case study villages, migrants from the hills called *Pahadis*, traditionally exacted from the original settlers called *Tharus*, unpaid labor for the construction and maintenance of village infrastructure, such as irrigation canals and roads. *Pahadis* usually enunciated the tradition as the system through which they had been introducing development into the backwater inhabited by uneducated *Tharus*. "People's participation", *Pahadis* argued, was an important tool not only to put the limited financial resources to maximum use, but also to nurture a sense of obligation among *Tharus* regarding the development of the villages.

The second level of "quadri-hermeneutics", drawing on interpretivism, clarifies more profound meanings than are immediately conveyed by the above analysis. The people studied are not mere "respondents" who unselfishly provide information for the benefit of the interviewer, but "reciprocators" who actively make out their cases.[22] At the time of this research, *Tharus* were increasingly expressing in public their discontent at the corvée tradition. The transition to a multi-party system in 1990, resulting in the relaxation of the political restrictions hindering freedom of speech and association, had provided the disadvantaged with a wider latitude in raising their grievances. Such societal forces building towards further social inclusion, therefore, rendered village leaders more defensive in the face of the researcher, whom they might understandably have suspected of sympathizing with an emerging group of NGOs voicing concern for the plight of *Tharus*. It can therefore be surmised that *Pahadis* drew on the rhetoric of "people's participation" as a convenient euphemism to evoke an impression of their commitment to public welfare, and thus to uphold their historical dominance in their villages.

The fourth level of the "reflexive" fieldwork methodology, namely self-reference through a post-modernist critique, leads us to a more nuanced understanding of *Pahadi-Tharu* relations. Researchers are liable to interpret interviewees' remarks as coherent and stable, disregarding the multiplicity and plasticity of meanings that the person attaches to his or her surroundings.[23] Post-modernistic thinking allows analysts to break out of this "realistic" mode, to pay attention to the transient and even fragmented nature of subject positions, which emanates from the contingent nature of discursive structures, or what Laclau and Mouffe term a "surplus of meaning."[24] One's articulatory practice cannot impose order with finality, but is penetrated by competing forces. Such precariousness is attested to by *Pahadis'* assertion of the corvée tradition, which entailed incoherent elements, reducing *Tharus* to underlings of *Pahadis*, while depicting *Pahadis* as "benevolent benefactors." The reasoning thus compelled *Pahadis* to face the dilemma of having to work to relax the corvée practice as "benign patrons."

Detailed empirical research on the implementation modality of village public works revealed that village leaders wavered between their propensity to maintain the corvée tradition, and their need to cater to public pressure to do away with the inequitable practice. As a result, at times, *Pahadis* even opted not to demand unpaid contributions, but to provide remuneration for *Tharus*. It was the post-modernistic approach, to focus on the "undecidability" of people's subjectivities, that brought to light halfway through the fieldwork my undue preoccupation with the binary rivalries between *Pahadis* and *Tharus*. I had treated the relations between the two, as if the former merely sought to exploit the latter's unpaid labor. I thereafter came to pay more attention to the ceaseless readjustments made to the ongoing social interactions between the two.[25] People's actual practices are not consistent but are contingent on specific situations or, as pointed out by Bourdieu, are "immersed in the current of time."[26]

Finally, the third level of "quadri-hermeneutics", building on the critical theory tradition, helped further delve into the subtlety of the power dynamics at work in the case study villages. Just as with the post-modernist approach, critical inquiry plays down the claim to "objectivity" made by proponents of "naturalism" that it is possible to accurately grasp "reality" through intimate contacts with the people studied. This is because people's and researchers' understandings of social phenomena are inherently distorted by macro socio-political structures that inconspicuously foster "false consciousness." A case in point in the context of this research is the "developmentalism" prevailing in Nepal, which is assimilated into the ways people see themselves and relate to others, to produce the construct of "villagers" who are in need of outside help to break with their "backwardness."[27] Moreover, most of those studying Nepal, with their moral commitment to rural "development", are prone to overemphasize the tyrannical nature of socio-political structures in the "periphery", and the resultant necessity to make external interventions to "emancipate" the disadvantaged in "outlying" areas.

Critical reflections on the impact of "developmentalism" on my research enabled me to realize that I had unknowingly constructed *Tharus* as being downtrodden by the corvée tradition. Close examinations of the on-the-ground situations showed that *Tharus* did not resign themselves to it, but offered both overt and covert resistance to the corvée norm, while complying with the general terms of labor obligations. The inconsistency in *Pahadis*' dealings with *Tharus*, stated above, was also attributable to the latter's resistance to the former's historical dominance. Without a critical inquiry into "developmentalism", I would have failed to take note of the leverage that *Tharus* were exercising to ameliorate the inequitable social order.

The interpretivistic, post-modernistic, and critical deliberations, thus far elaborated, which correspond to the last three facets of "quadri-hermeneutics", do not diminish the importance of the first level of "naturalistic" inquiries, intended to allow the interviewees to "speak their minds" in "non-artificial" settings. To grasp the multiple and transient nature of the "reality", it was essential that I engaged in "naturalistic" reflections to monitor the influence of a particular research setting on the interviewees, and attempted to create an atmosphere in which they felt at ease to divulge stories that would otherwise have remained untold.[28] It was imperative, at the same time, not to romanticize the interview-interviewee interaction as a joint construction of "determinate" meanings, and to avoid lapsing into "naive realism." I instead sought to grasp how "(m)eaning and understanding shift . . . across time, and across situations"[29] during my close and long-term contact with the two case study villages.

Such multiple reflections, as exemplified in this section drawing on my village-level fieldwork, were also pursued in interacting with "non-villagers", that is government officials and political leaders at the district and national levels. As described from chapter 3 onwards, "developmentalism" set the overall tone of the unfolding of "participatory" river control, for which the policy decision to promote "participation" was enforced irrespective of location-specific circumstances. At the same time, district and national-level policymakers did not unambiguously seek to impose "participation" in all local river control projects. On the contrary, just as *Pahadis*, who wavered between their propensity to exact corvée, and their need to relax the inequitable tradition, policymakers also held diverse, multifaceted subject positions. "Quadri-hermeneutics" has been as indispensable in delving into the contingent nature of policymaking, as it has been vital for grasping the uncertainty of village power dynamics.

## Overview of the Book

The book consists of seven chapters, beginning with this introduction. Chapter 2 provides the theoretical framework for the study, while chapters 3 to 6 present the analyses and findings of the empirical data that I gathered during my field research. In chapters 3 to 6, I trace the unfolding of the "participatory" river control policy at the national, district, and village levels. The concluding chapter brings together major empirical findings, as well as theoretical and practical implications.

Chapter 2 examines existing literature to provide the analytical framework for this book. It begins with an examination of the Foucauldian view and Giddens' "structuration" notion of power, both of which highlight the meshing of domination and resistance, to draw attention to the feature overlooked by the conventional "power as domination" view, which divides society into those who have power and those who do not. The domination/resistance nexus opens up space for ordinary people to react to and resist tyrannical forces inherent in the policy process. In the light of the leverage that the general public potentially exercises over the policy process, the chapter also makes a critical assessment of existing research on public participation in policymaking. It suggests moving away from a long-standing tradition in development studies to presume the need to launch a distinctive set of countermeasures to emancipate people from the "exclusionary" policy process. The final section of this chapter deliberates on how best to analyze policymakers' discursive practices, underlying the contested nature of policymaking that entails constant reflections and readjustments.

Chapter 3 examines how the country's decentralization and participation (D&P) policy that held sway in the 1990s spilt over to the "participatory" river control policy at the center, in such a manner as to relegate flood mitigation to the delivery of pre-packaged programs to flood-affected areas. The D&P rhetoric embodied in the policy gave an illusion of overhauling the conventional, top-down conduct of development interventions. While the overall tone of "participatory" river control was set in this way, at the same time, central policy-makers did not seek to unilaterally impose flood mitigation packages. They instead felt ambivalent about the blueprint embodied in the policy, given the vulnerability of the policy to counter-discourses that upheld more emancipative approaches. As a result, the manner "participatory" river control was translated into action was ceaselessly reflected upon among central stakeholders, who were caught in a dilemma over whether or not to relegate the policy to such a centrally directed process. In the latter half of the 1990s, the central government came to ignore the "participatory" river control policy. Chapter 3 concludes by explaining the downward trend of the policy, in the light of the contingent and inconsistent subject positions of policymakers, and the resultantly contested and fluid nature of the policy process.

This "ascending" approach to the "participatory" river control policy, that is the examination of the unfolding of the policy as a web of micro-power struggles, is also taken in chapters 4, 5 and 6 that deal with local-level studies of Bardiya. Chapter 4 first analyzes how district-level policymakers interpreted and contested the meaning of the "participatory" river control policy. The "participatory" river control policy was assimilated into the local corvée tradition by local politicians and government officials, who appropriated the policy directive to justify the age-old, hierarchical social order. At the same time, just as in the case of the central-level policy process taken up in the preceding chapter, policymaking in Bardiya entailed ceaseless renegotiations among local leaders who wavered between the corvée-based construction of physical structures, and the need to do away with the unjust practice. The chapter seeks to explicate the unstable and incoherent subject positions of district-level stakeholders, in the light of the increasing penetration into their discursive fields, of such counter-discourses as those condemning the injustice embodied in the corvée tradition, or problematizing the constrictive view that equated flood mitigation with the construction of physical structures.

The continual struggles over the policy's localized meanings at the district level were also a reflection of the contested nature of grassroots power dynamics, with which the unjust social order was constantly renegotiated. Chapters 5 and 6 take up two villages, to assess how the "participatory" river control policy unfolded at the community level. As is surmised from the notions of power, taken up in chapter 2, which point to the intermingling of domination and resistance in daily social interactions, villagers did not passively accept, but ceaselessly contested the way the policy was put into action. Furthermore, just as leaders at the district level, village elites in the two areas did not simply seek to preempt the "participatory" processes, but were caught in a dilemma as to whether or not to promote the corvée-based river control. In the light of the complex meshing of varying meanings attached by different villagers, both chapters examine to what extent and how villagers exercise influence over the policy process, including the decline of the river control policy towards the end of the 1990s.

Chapter 7 concludes this book by summarizing the empirical findings, and by extracting key theoretical and practical considerations. The "ascending" approach of this study, examining the micro-power struggles experienced by various stakeholders, illustrates the existence of a fertile ground for policy "clients" to exert leverage over the unfolding of the river control policy, and subsequently the nationwide downturn in the policy. I then deliberate the theoretical implications for research into people's participation in policymaking, which has tended to simplistically take for granted the necessity of external interventions to turn the general public into "citizens" capable of engaging in policymaking. The chapter ends with a suggestion that proponents of greater public participation in the policy process should pay attention to the emancipatory potential arising from people's everyday "cultural politics", that is the day-to-day flow of their engagement with policymaking. It is then possible to conceptualize strategies that build on, and compensate for the limitations of ongoing local power struggles to overcome entrenched inequalities.

## Notes

1. Excerpts from my interview of July 2002.
2. Excerpts from my interview of July 2002.
3. John Gaventa and Camillo Valderrama, *Participation, Citizenship and Local Governance: Background Paper for Workshop, Strengthening Participation in Local Governance* (Brighton: IDS, 1999), 1-6; Andrea Cornwall and John Gaventa, *From Users and Choosers to Makers and Shapers: Repositioning Participation in Social Policy* (Brighton: IDS, 2001), 9-19.
4. Andrea Cornwall, *Beneficiary, Consumer, Citizen: Perspectives on Participation for Poverty Reduction* (Stockholm: SIDA, 2000), 61-62.
5. John Gaventa, "Introduction: Exploring Citizenship, Participation and Accountability," *IDS Bulletin* 33, no. 2 (2002): 3-6.
6. For example, see Cornwall and Gaventa, *Users and Choosers*; Sam Hickey and Giles Mohan, eds., *Participation, from Tyranny to Transformation?: Exploring New Approaches to Participation in Development* (London: Zed Books, 2004b).
7. Sam Hickey and Giles Mohan, "Towards Participation as Transformation: Critical Themes and Challenges," in *Participation, from Tyranny to Transformation?: Exploring New Approaches to Participation in Development*, eds. Sam Hickey and Giles Mohan (London: Zed Books, 2004a), 3-24.
8. Andrea Cornwall, *Making Spaces, Changing Places: Situating Participation in Development* (Brighton: IDS, 2002), 10-17.

9. Michel Foucault, *Discipline and Punish: The Birth of the Prison* (London: Penguin Books, 1976).

10. Michel Foucault, "Afterword: The Subject and Power," in *Michel Foucault: Beyond Structuralism and Hermeneutics*, ed. Hubert L. Dreyfus and Paul Rabinow (Brighton: The Harvester Press, 1982), 220.

11. Cris Shore and Susan Wright, eds., *Anthropology of Policy: Critical Perspectives on Governance and Power* (London: Routledge, 1997).

12. Among earlier noted proponents are Escobar (1995), Ferguson (1990), and Pigg (1992).

13. Arturo Escobar, "Beyond the Search for a Paradigm: Post-Development and Beyond," *Development* 43, no. 4 (2000), 14. Emphasis added by this author.

14. Michel Foucault, "Two Lectures," in *Power/Knowledge: Selected Interviews and Other Writings 1972-1977 Michel Foucault*, ed. Colin Gordon (New York: Harvester Wheatsheaf, 1980), 99-100.

15. Foucault, "Two Lectures," 99-100.

16. There exists some studies, such as those by Schaffer (1984) and Long (1992) that similarly analyzes policies by heeding specific situations that lead to particular "practices" instead of deducing the policy process abstractly from the structural locations of policy stakeholders. At the same time, those studies tend to omit certain groups from their analyses of the autonomy and spontaneity entailed in people's daily practices. In several studies, "target groups" are assumed to have little leverage over the policy process, while in others, high-level policymakers are conceived as actors essentially prone to impose on local people policies that are incompatible with their life worlds. I pursue a rather different approach by taking a holistic look at various actors at different levels, by tracing the emergence and evolution of the "participatory" policy in the capital, as well as its local manifestations at district and village levels.

17. In Pyauli, a sample of households was chosen from the voter list. For this, I first grouped voters by settlements, and used a random number table to select a sample from each settlement. On the other hand, in Majuwa, all the households were surveyed. A total of approximately 690 surveys was undertaken, a sample of about 140 households in Majuwa, and about 550 in Pyauli. The main objective was to gather data on the ways in which each household made labor contributions. The questionnaire thus included questions, such as "who participated for how many days", "how many hours the participants worked", "who told them to volunteer for the projects", and "with whom the participants went to the project site."

18. Martyn Hammersley and Paul Atkins, *Ethnography: Principles in Practice*, (London: Routledge, 1995), 11.

19. Bending back on oneself drawing on "naturalism" is not "reflexive" but "reflective" in that it focuses only on other persons' constructions of reality, not that of the researcher. On the other hand, the "reflexive" methodology I pursued takes into account the broader socio-historical context in which the villagers and I co-constructed research outcomes, as described in this section. For discussions on the distinction between "reflective" and "reflexive", see Mats Alvesson and Kaj Skoldberg, *Reflexive Methodology: New Vistas for Qualitative Research* (London: Sage, 2000); Frederick Steier, "Introduction: Research as Self-Reflexivity, Self-Reflexivity as Social Process," in *Research and Reflexivity*, ed. Steier Frederick (London: Sage, 1991a), 1-11.

20. Charlotte A. Davies, *Reflexive Ethnography: A Guide to Researching Selves and Others* (London: Routledge, 1999), 7.

21. Alvesson and Skoldberg, *Reflexive Methodology*.

22. Frederick Steier, "Reflexivity and Methodology: An Ecological Constructionism." in *Research and Reflexivity*, ed. Frederick Steier (London: Sage, 1991b), 163-85.

23. Jonathan Potter and Margaret Wetherell, *Discourse and Social Psychology: Beyond Attitudes and Behaviour* (London: Sage, 1987), 34.

24. Ernesto Laclau and Chantal Mouffe, *Hegemony and Socialist Strategy*. 2d ed. (London: Verso, 2001), 111.

25. As shown by this episode, researchers are not necessarily in a privileged position to represent social phenomena, because they are also bound by societal norms and thus tend to produce discourses that add to taken-for-granted standards. Since, as discussed by Foucault (1976), "knowledge" production is not immune from such "disciplinary" power which subjugates people to common beliefs, "social scientists are very much in the discourse distribution and reinforcement business" to use the phrase found in Alvesson's work (2002, 140). It is, therefore, crucial for researchers to reflect on the language they use in summarizing research findings, with a view to unsettling ostensibly "factual" accounts manufactured through text production.

26. Pierre Bourdieu, *The Logic of Practice* (London: Polity Press, 1990), 81.

27. Stacy L Pigg, "Unintended Consequences: The Ideological Impact of Development in Nepal," *South Asia Bulletin* 13, no. 1&2 (1993): 45-58.

28. It was usually during casual "interviews" that villagers provided clues to comprehending the transient and multi-faceted nature of individuals' identities. For example, while walking home after an interview in which he adamantly upheld "people's participation", one village leader admitted to me the need to ease up on the corvée practice in view of the growing resentment about it. A peasant, who had attributed the corvée tradition to the asymmetrical social relations in an interview, expressed to me his feeling of ambivalence, while having tea, about the use of contractors who are liable to cut corners. Engaging in unobtrusive, less formulaic interactions, as is advocated in "naturalism", therefore, proved crucial in generating insights into the uncertainty of various local actors' identities.

29. James J. Scheurich, *Research Method in the Postmodern* (London: Routledge/ Falmer, 1997), 62.

# Chapter 2
# Theoretical and Conceptual Frameworks: Power, Participation, and Policy

This chapter provides theoretical and conceptual frameworks for this book by reviewing existing literature on the nature of power, on popular participation in the policy process, and on discourse analysis. As explained in chapter 1, the main concern of this study is to look into the interfaces of the "official" policy discourse and local actors in the evolution of "participatory" river control in Nepal, and to examine whether policy "clients" exercise some leverage over the unfolding of policymaking. For this purpose, it is imperative to develop analytical frameworks, both to delve into various facets of power dynamics, and also to make sense of their implications for the policy process and for the discursive practices of policymakers.

This chapter begins by analyzing two strands of the debates on power, namely the "structuration" perspective, and the Foucauldian notion, both of which refuse to dichotomize society neatly into the "powerful" and the "powerless." The "structuration" notion regards power as immanent in social interactions, through which various actors, both "powerful" and "powerless", respond to both the opportunities and constraints arising from daily routines. According to the Foucauldian view, "disciplinary" power pressurizes people to conform to social norms, and at the same time facilitates various actors to renegotiate the definitions of these societal standards. The two conceptions of power thus illustrate the complex meshing of domination and resistance in day-to-day lives, and help us to arrive at an in-depth understanding of power dynamics.

This chapter then considers the implications of such intricate workings of power on the policy process. By drawing on the "structuration" view of power as constantly contested by social actors, the "actor-oriented" approach explains how policy interventions are renegotiated and reworked at local levels. The "anthropology of policy", on the other hand, describes how "disciplinary" power subjects the policy process to "cultural politics", through which stakeholders, varying from policymakers to their clients, contest with one another over policy direction. This section then builds on these studies, to deliberate whether public engagement with the policy process needs be activated through external interventions, or whether it is already immanent in day-to-day policymaking endeavors, given "cultural politics" that are played out among various players.

This book thus questions a long-standing tradition in studies on the policy process to equate it with a top-down sequence of taking decisions and implementing them. This chapter concludes by laying out a framework for discourse analysis, in view of the central theme of this study to correct the conventional bias toward the "logical" and purposive nature of policymaking, and instead, to highlight the uncertainty entailed in the process. As stated towards the end of this chapter, the discursive structure of a policy directive is inherently vulnerable to other competing interpretations because it attempts to impose order by deliberately excluding other alternative options. The unfolding of policymaking, however imposing and exclusionary it may seem, is therefore a contested process subject to constant reflections and readjustments.

# Oppression/Resistance Nexus in Power Dynamics

Debates of power among political theorists traditionally revolved around where to draw the line between one's autonomous actions and thoughts, and those distorted by other actors.[1] The "orthodox" accounts of power originated from Dahl,[2] who considered it to exist when A shapes decisions that compel B to do what s/he would not do otherwise. One of the most notable criticisms was raised by Bachrach and Baratz,[3] who pointed out that power also manifests itself when A molds the political agenda in such a manner as to prevent B's concerns from arising. Lukes,[4] in his seminal work, *Power: A Radical Review*, subsequently contributed to the enrichment of the "mainstream" debates by bringing to light the third dimension of less visible "thought control", emanating from A's manipulation of what B thinks and wants.

Lukes[5] concedes, in his second edition of *Power: A Radical Review*, several shortcomings of this so-called "three-dimensional" debate. First, it was committed to the "exercise fallacy" that power is present only when A exerts power over B, although power, when viewed in terms of agents' abilities, may not be and need not be exercised in order to effect outcomes. Moreover, it focused narrowly on "domination" under which A harms B's wants and needs, while neglecting "beneficent power", which A can use to satisfy and advance B's interests. Finally, it reduced societal dynamics to binary power relations between actors with unitary interests, thus failing to consider the multiple and often incoherent nature of identities that renders it infeasible to divide members of society simplistically into "goodies" and "baddies." Power thus conceived in the "orthodox" debate (hereinafter called the "power as domination" view, to draw on the term used by Lukes[6]) was a partial account of the topic, in that it presented only the "power-over" perception that pitted resistance by "oppressed" groups against control by "dominant" players.

This section takes up two strands of the debates on power that shed light on the "entanglements" of oppression and resistance,[7] namely the Foucauldian perspective and Giddens' "structuration" view. Both theories consider that society is is not neatly grouped into the "powerful" and the "powerless", and thus reject the binary opposition between oppression and resistance. On the contrary, according to Foucault, power exerts pressure on people both to conform to, and to resist societal norms, thereby facilitating renegotiations of prevailing social standards. The "structuration" notion regards social actors as responding to the limitations and opportunities that emerge in their daily routines, and as a consequence, interpersonal social relations are ceaselessly renegotiated and modified. The two perceptions of power thus bring to light the constant remolding of societal power dynamics, which enables us to gain a more nuanced understanding of the nature of power than does the traditional "power as domination" view that represents power contestation in static terms.

## "Disciplinary" Power

The alternative notion of power advanced by Foucault marks a radical departure from the conventional "power as domination" perception, in that it does not ascribe power to any particular actor. Foucault considers power to be omnipresent and to permeate society in a manner such that everybody is affected by it. According to Foucault, power is intrinsically related to "knowledge." What is regarded as "knowledge" upholds certain social norms and standards, which are assimilated into the way individuals see themselves and relate to others. Foucault terms this indirect, insidious form of social control "disciplinary" power.[8]

According to Foucault, "disciplinary" power is both "totalizing" and "individualizing."[9] It produces "totalizing" effects which exert pressure on people to conform to prevailing norms, and to marginalize those who fall outside the "acceptable" range. It is also "individualizing" because human beings are classified and compartmentalized in accordance with the extent to which each of them lives up to social standards. The central tenet of "disciplinary" power is that it acts on individuals at the micro level, whose behaviors are molded to forge "docile bodies." As a result, individuals discipline themselves through self-monitoring and self-measurement of their compliance with social norms. In this way, the human body serves both as the object of, and as the instrument for "disciplinary" power.

However, it is impossible for "disciplinary" power to forge complete compliance with social norms and standards. Different individuals or groups attribute varying meanings to particular situations, and constantly struggle with one another over the definition of norms and standards. Social actors therefore contribute, through their daily discursive practices of making sense of social phenomena, to renegotiating and reshaping the "disciplinary" power that defines social norms. In the words of Foucault, "(t)he individual is an effect of power and at the same time . . . it is the element of its articulation."[10] Foucault argues that an individual is not only a passive vehicle of "disciplinary" power, but also an agency to reproduce and modify prevailing discourses, although he tends to downplay a discoursing subject.[11] According to Foucault, therefore, "there is no such thing as power as a whole" but power is exercised "in diverse and multiple ways at the 'micro-level.'"[12] Power circulates in the form of a chain, and can be grasped by mapping out the web of power struggles in which social actors are engaged in myriad places.

What renders micro-level power contestation even more dynamic is the provisional and contingent nature of actors' identities. Subjectivity, that is how individuals understand themselves and others, is constantly in the process of being shaped through the discursive practices with which different social actors continually negotiate and renegotiate their interpretations of reality. Identity is thereby far from being fixed but is constantly reproduced and transformed through one's daily social interactions. Given this fluid nature of subjectivity, how particular actors react to "disciplinary" power cannot be deduced a priori from their presumable interests. On the contrary, when and where "disciplinary" power is in operation, a "whole field of responses, reactions, results, and possible inventions may open up."[13]

The starting point of power analysis should therefore be to focus on the "antagonisms of strategies",[14] that is on how different actors in particular situations are subjugated to and resist "disciplinary" power. One should conduct an "ascending" analysis that starts by examining the micro-power experienced by different individuals and groups, and then try to reveal a more general, overarching power dynamics.[15] A conventional "descending" approach would be to investigate which individuals or organizations possess power and what strategies they pursue in order to exercise power over others. Rather than assessing power from the angle of institutions, however, it is imperative to "analyze institutions from the standpoint of power relations."[16] As mentioned above, power does not fall into the hands of particular agents, but permeates and circulates in society. One must instead pay attention to what Foucault terms the "micro-physics of power"[17] to analyze what effects "disciplinary" power exerts on particular individuals or groups. According to Foucault, therefore, power should be examined as a regime of micro-level, inconspicuous practices that different actors engage in, in the face of "disciplinary" mechanisms.[18]

According to Foucault, power permeates in society to bring all social actors under its sway, and is not attributed to any particular agent. The Foucauldian notion is therefore criticized for its disregard of human agency to counteract the influence of power. However, this does not mean that it is incompatible with the conventional perspective that entrusts power to actors.[19] Despite their inability to stay immune from "disciplinary" mechanisms, human beings are capable of sidestepping the grasp of power, and are never tamed completely, as pointed out by Foucault. "Disciplinary" power not only shapes one's identity, but also enables each individual to position himself or herself in the light of prevailing social norms and standards, whether compliantly or not.[20] Power is not merely oppressive, but is also productive in the sense that it generates "knowledge" and thereby facilitates social actors to establish their standpoints and pursue their goals. According to Foucault, therefore, power dynamics are both non-subjective as well as intentional.[21]

The Foucauldian perception and the more mainstream approach of attributing power to particular actors complement each other, in that the former is useful in clarifying the dynamics that forge the identities and norms underlying interactions among different social actors.[22] In turn, the latter operates back on the former in that interpersonal power contestation serves to reproduce and modify the "disciplinary" power that upholds social norms. As quoted above, "(t)he individual is an effect of power and at the same time . . . it is the element of its articulation."[23] In view of such mutuality between the two seemingly disparate perspectives, this study draws on both the Foucauldian notion, and Giddens' "structuration" perception of power, taken up in the following section, that ascribes power to particular players.

## "Structuration" View of Power

Although the "structuration" perception, proposed by Giddens,[24] attributes power to particular agents just as does the "power as domination" perception, it is similar to the Foucauldian notion in regarding power not only as oppressive, but also as productive, in that it facilitates social actors to form and pursue their agenda. In this respect, the "structuration" notion sheds light on the dialectics between individual actions and social structure. In the mainstream "power as domination" debate, the agency-structure tension has remained unresolved,[25] as attested to by a remark by Lukes, who attempted to assimilate structural determinism into the "power as domination" perception, but admitted the difficulty of determining "where structural determination ends and power and responsibility begin."[26]

In order to overcome such dualism between agency and structure, Giddens puts forth the notion of the "duality of structure." According to Giddens' "structuration" theory, social structure is both constituted by human agency, and at the same time, serves as a condition for this constitution. Structure constrains individuals' thoughts and actions, and enables human agents to draw on "structural properties"[27] in accommodating themselves to, and challenging regularized patterns of social interactions. Underlying such an agency/structure nexus is the "reflexive form of the knowledgeability of human agents",[28] albeit bounded by structural elements of society. Individuals often know, as a second nature, about what works and how, which Giddens terms "practical consciousness." Giddens' notion of "practical consciousness" corresponds to Bourdieu's concept of "habitus"[29] which operates at the subconscious level, and provides agents with an intuitive sense of how to act and react in their daily lives. Moreover, according to

Giddens, it is also possible for individuals to dwell discursively on the purposes of their actions, drawing upon their "discursive consciousness."

Either unknowingly through "practical consciousness", or reflexively using "discursive consciousness", social actors monitor actions of their own and others, as well as the contexts in which daily conduct takes place. As a consequence, human agents do not simply pursue their routines, but also strive to adjust their daily activities in order to turn "structural properties" to advantage. In this way, human agents continually contribute to the production and reproduction of social structure. As pointed out by Scott, even under the most coercive environments, subordinate groups engage in low-profile "infra-politics" that serve, albeit in disguised, undeclared forms, not just as building blocks for open challenge, but in themselves constitute "practices that aim at an unobtrusive renegotiation of power relations."[30]

According to Giddens, power is founded on this "dual" nature of human agency that constitutes, and is also conditioned by "structural properties."[31] Power is "the transformative capacity"[32] to intervene in, and affect the routinized flow of social interactions by going through the above reflexive process. Power is therefore not one category of human conduct, nor is it a resource for powerful A to mobilize to influence powerless B. It is "instantiated in action, as a regular and routine phenomenon."[33] While power is conventionally considered to entail coercion and conflicts, Giddens regards power as an integral element of all social interaction. As described later in chapters 5 and 6 on the village case studies, it is not always plausible to pinpoint clashes of interests, overt or latent, given the provisional and multifarious nature of individuals' subject positions. One advantage of Gidden's notion of power is that it allows us to focus our attention on the effects of power, without having to impute interests to social actors, or to judge whether B's interest is harmed or not.

Another strength of Giddens' view is that it elucidates the meshing of domination and resistance. As illustrated in chapters 5 and 6, even when actors submissively conform to prevailing social norms, human interactions result in constant readjustments to existing social relations. Such intermingling of oppression and rebellion is facilitated by covert resistance by disadvantaged groups, or what Scott terms "infra-politics", as referred to above. What makes "infra-politics" subtle is that "[m]ost acts of power from below even when they are protests . . . will largely observe the 'rules' even when their objective is to undermine them."[34] In the face of such low-profile but down-to-earth resistance by disadvantaged groups, power holders are compelled to work ceaselessly to maintain their domination and control. The agency/structure nexus described by Giddens helps capture such ceaseless reproduction and modifications of "structural properties" in the day-to-day flow of events. Domination and resistance are so interwoven that it is implausible to analyze them separately.

Giddens argues that power is conventionally reduced either to actors' choices, as described by Lukes, or to properties permeating society, as argued by Foucault.[35] According to Giddens, the "structuration" notion of power is instrumental in bridging this gap between voluntarism and determinism, and in analyzing the interactions between the two aspects from the viewpoint of the "duality of structure." At the same time, as stated in the preceding section, Foucault argues that an individual is not only a passive vehicle of "disciplinary" power, but also an agency to reproduce and modify prevailing discourses, although he tends to downplay a discoursing subject. The above statement by Giddens therefore does not hold true in that the Foucauldian notion also establishes a nexus between voluntarism and determinism, by regarding power as both non-subjective as well as intentional.[36]

This study brings together the Foucauldian and the "structuration" views, capitalizing on their common orientation to regard domination and resistance to be immanent in daily social interactions. The "structuration" perspective elucidates how social actors respond to both the limitations and the opportunities that emerge in the day-to-day flow of their lives, and ceaselessly renegotiate ongoing social relations. The Foucauldian conception, on the other hand, brings to light how "disciplinary" power that exerts pressure on people to conform to pervasive norms compels various actors to struggle with one another over the definition of these societal standards. Both the dynamic nature of interpersonal social relations, highlighted by Giddens, and the ceaseless adjustments to societal standards, pointed out by Foucault, provide fertile soils for one another. By combining the two facets of power, it is possible to pay attention to the synergy between them, and thus to arrive at an in-depth understanding of power dynamics.

## Ongoing People's Engagement with Policymaking

Building upon the theoretical frameworks of power, described above, this section reviews the existing literature on the policy process. The Foucauldian notion of power and Giddens' "structuration" view, both of which shed light on the interweaving of domination and resistance in daily social interactions, are useful in making sense of how the policy process continues to be remolded and remade by ongoing power struggles. On the other hand, a lasting trend in policy studies is to focus on elucidating the forces of marginalization at work in the policy process. This is attributable to a commitment shared by many researchers to take the side of ordinary people, and so to overemphasize the oppressive nature of power at work in the policy process. For example, according to one strand of policy research, which Parsons terms "argumentative" studies,[37] the policy process is prone to be controlled by politicians, bureaucrats, and specialists, who are inclined to define social problems in such a manner as to entrench their clients' dependence. In order to uncover the implicit values lurking in official policy narratives, "argumentative" studies suggest such countermeasures as "counter-labeling"[38] and "frame reflection"[39] through which the general public come to pose a more principled challenge to the manipulative process of policymaking.

However, as pointed out by Hill with reference to Giddens' "structuration" theory, the policy process is not a simple reflection of elites' interests.[40] The literature review of existing policy studies, undertaken in this section, is intended to arrive at a more nuanced understanding of the power dynamics at work in policymaking, in line with the intricate nature of power elucidated above. The two strands of policy studies, desribed below, follow the "structuration" view and the Foucauldian notion of power, and shed light on the intricate meshing of domination and resistance in the policy process. As discussed below, such a polarized view of elite control and public resistance, as is concieved under the "argumentative" approach, overestimates the perfidies of policymakers, and also fails to take note of the leverage that ordinary people can exercise over the policy process.

I then turn attention to development studies where there are increasing calls for public engagement with the policy process. This group of research is similarly liable to consider that ordinary people are "barred" from the policy process, and propose a distinct set of activities to activate public engagement in policymaking. Contrary to the prevailing assumption about the need to initiate a separate set of correctives, I would argue that, even under "dictatorial" circumstances, the policy process continues to be reshaped and remade by ongoing power struggles, through which ordinary people influence the way policy is initiated and carried out.

## "Actor-oriented" Approach

This section examines the "actor-oriented" approach which was promulgated by Norman Long and his colleagues at Wageningen in the Netherlands. The "actor-oriented" view provides a useful insight into how Giddens' "structuration" theory manifests itself in the policy process. According to the "actor-oriented" approach, policy studies require more than an analysis of external interventions per se, and should also examine ongoing local power dynamics. An external intervention is implanted in a locality where various social actors are already pursuing their own "projects" to enunciate or challenge existing social order. External interventions take on different meanings to various individuals and groups. Policy therefore needs be "deconstructed" in view of how various local actors assimilate or challenge "official" policy narratives, in the context of local power struggles.[41] The "actor-oriented" analyzes the policy process "for what it is – an ongoing socially-constructed, negotiated, experiential and meaning-creating process" at the micro-level.[42]

The "actor-oriented" approach not only argues that external interventions are not executed in a linear, step-by-step manner. It goes further to recognize the potentials of "target groups" to influence planned interventions, by drawing upon what Giddens terms the "knowledgeability of human agency", that "attributes to individual actors the capacity to process social experience and to devise ways of coping with life, even under the most extreme forms of coercion."[43] "Target groups" therefore do not passively accept policy directives from above, but exercised their human agency to rework them to suit location-specific social, political, cultural circumstances. How a policy intervention unfolds on the ground should therefore be induced from empirical investigations of how various state and societal actors interact in specific local contexts.[44]

There exist studies, other than those drawing on the "actor-oriented" view, that do not subscribe to the prevailing tendency to overestimate the perfidies of policymakers. The "actor-oriented" view provides a new twist to those studies by highlighting the latitude for policy "clients" to respond to the onslaught of official discourses.[45] For example, "policy network" analysts similarly illustrate how state and non-state stakeholders make formal or informal contacts and bargain over policy goals and strategies, to highlight the diverse patterns of state-society representations in the policy process.[46] However, the scope of "policy network" studies fails to encompass grassroots interfaces, unlike the "actor-oriented" approach. Studies on "room for maneuver", associated with the works of Clay and Schaffer, and of Grindle and Thomas,[47] point out the existence of "policy space", in which high-level politicians and officials can introduce imaginative endeavors even under unfavorable circumstances. Lipsky focuses on how "street-level government workers" use their discretion in shaping the policy process.[48] The "actor-oriented" approach is seminal in the sense that it further moves forward the debate to encompass "target groups" at the community level.

As pointed out by Arce, Long, and others,[49] the "actor-oriented" view is not confined to an analysis of face-to-face interactions, but is also concerned with "how 'external' or geographically distanced actors, contexts and institutional frames shape social processes, strategies and actions in localised settings."[50] At the same time, it overestimates the capacity of policy clients to absorb and rework extra-local initiatives, and fails to consider "broader cultural models or organizing principles, movements and changes" that are outside the scope of such interface analysis.[51] This drawback stems from a pitfall of Giddens' "structuration" notion, which, according to Clegg,[52] attempts to elaborate the nexus between structure and agency, but patches up the tension by according agency a predominant role over structure. This unresolved dualism of the "structuration" view manifests itself in the "actor-oriented" approach. By simplistically arguing that local actors are not subalterns subjugated to larger societal forces but are capable of translating them into localized meanings and practices, the "actor-oriented" approach disregards such overarching socio-cultural dynamics.

Moreover, as acknowledged by Arce,[53] another strand of criticism is that "actor-oriented" studies leave out "policy elites" from their analyses, while restricting their scope of analysis to micro-level interactions. This "localism" results in a blind spot where the "actor-oriented" approach fails to consider the potentiality of policy "clients" to influence higher-level policymaking. It does not illuminate how and to what extent beneficiaries exercise leverage over the general, overarching evolution of the policy process. Another related weakness is that, without making empirical investigations into their actual practices, it accords high-level policymakers a fixed disposition to impose programs incongruent with local circumstances. From the viewpoint of "actor-oriented" analysts, such criticisms against local orientation are "missing the point"[54] because the main focus is to elucidate the inability of top-down policy directions to unfold on the ground as originally intended by policymakers.

At the same time, the "actor-oriented" approach is neither static or monolithic, as illustrated by the defense made by Norman Long in a seminar, "Yes, but that Norman Long does not exist any more."[55] Long accordingly, in some of his recent works, refers to the need to take into account the fragmentation and multiplicity of the intents of policymakers,[56] as well as the broader socio-cultural paradigms that shape the life-worlds of policy clients.[57] However, such overarching societal dynamics, in which negotiations among actors at multiple levels take place, fall outside the purview of "actor-oriented" analyses.[58] Moreover, the approach does not dissect the non-linear, fragmented process of policy formation at higher levels, but focuses only on social practice situated at particular local sites.[59] Policy analysts should therefore seek other approaches in explicating these issues. The "anthropology of policy", taken up in the following section, sheds light on a wider socio-cultural paradigm, while discourse analysis, described in the final section of this chapter, highlights the contested and incoherent nature of high-level policymaking.

## "Anthropology of Policy"

The "anthropology of policy" intends to deconstruct the discursive processes by which policies are constructed.[60] Like the "argumentative" perspectives described earlier, this strand of debate considers that policy language manufactures social reality and solutions to its problems. The "anthropology of policy" further delves into how policy narratives pressurize ordinary people to embrace the expansion of state control in society, or what Foucault terms the "governmentalization" of society. This type of "disciplinary" power operating in the policy process subjugates "target groups" into passive, constrained positions, and at the same time, facilitates them to position themselves vis-à-vis "disciplinary" forces and to shape their counter-strategies. A clue to redressing the manipulative nature of the policy process, according to the "anthropology of policy", is the political activism of policy "clients", or "cultural politics" delineated later in this section, rather than another set of external interventions to redress the perfidiousness of policymaking.

According to Foucault,[61] the central focus of current governmental practices is the growth and care of population, rather than the protection of territorial sovereignty that was the preoccupation of pre-modern governments. This major shift was facilitated by the emergence of capitalism because "the accumulation of men and the accumulation of capital . . . cannot be separated."[62] Scientific categories concerning biological existence, such as birth rate, longevity, and public health, began to be placed on the political agenda, and the survival and welfare of human species was brought into the realm of governmental surveillance. The rise of "bio-politics" was accompanied by the emergence of new technologies of demographics to keep track of various traits of populations, and the rise of economics to analyze relations between wealth, production, and populations.

Foucault points out that the contemporary age is the "era of governmentality."[63] In the name of public welfare, various aspects of social life have become the focus of state interventions. As pointed out by Shore and Wright,[64] public policy came to prominence as an instrument to justify and legitimatize the expansion of state control, or the "governmentalization" of society. Policy functions as a "political technology", the term coined by Foucault,[65] to reduce a complex issue entailing conflicts of values and interests, to a technical matter of identifying an optimal solution, thus legitimizing the proliferation of state interventions. In the disguise of neutral, scientific language, "official policy" then advances the fallacy of what Schaffer terms "decisionality"[66] to treat "decision-making" as a rational, technical, bureaucratic process distinct from "implementation." A policy objective is therefore given the appearance of a foregone conclusion founded on a rigorous analysis of viable alternatives that has already been undertaken during the "decision-making" stage. As a result, "disciplinary" power is exercised to prompt individuals to unknowingly embrace the norms and values unwittingly upheld by the policy in question, which also produces the unintentional side effects of reinforcing the trend of "governmentality."

In the field of development studies, the "anthropology of policy" is adopted by "post-development" thinkers who regard development as a discourse that fabricates reality in such a manner as to embrace the "governmentalization" of society (among earlier noted proponents are Escobar, Ferguson, and Pigg).[67] A major criticism leveled against the "anthropology of policy" is that it presents an overgeneralized, essentialized view of development policies as singular, hegemonic and oppressive.[68] "The development discourse" is neither static nor monolithic and is increasingly challenged by those working both within and outside the industry. In reality, there exists a variety of top-down development policy initiatives, some of which may serve as an inspiration for their intended beneficiaries, rather than working to legitimatize structures of power and domination. These critics, while still sympathizing with critical discourse analysis entailed in the "anthropology of policy", propose not only to deconstruct development policies, but also to reconstruct them in such a manner as to fully unleash their progressive potentials.

Proponents of the "anthropology of policy", on the other hand, cast doubt on the pervasive centrality of development as an organizing principle of social change. Escobar refutes the above criticism by pointing out that its "chronic realism" lies in assuming that development is unanimously desired by the "people."[69] Its weakness lies in its "naturalized morality" that what is at stake is the livelihoods of marginal groups, and that policy analysts should strive to provide a clear model of how social change can be effected through development policy interventions. Protagonists of the "anthropology of policy" also take the side of "disadvantaged" groups, but advocate alternative strategies of engaging in the "utopian possibility of reconceiving and reconstructing the world from the perspective of, and along with, those subaltern groups that continue to enact a cultural politics of difference."[70] To use the phrase found in Moore's work, "[i]nstead of conceiving development as overdetermining the shape of politics", we should pay attention to a wider spectrum of "grounded livelihood struggles."[71]

Fagan's assertion that it is crucial to move away from "cultural analysis"[72] is useful in elucidating the "cultural politics" upheld by the "anthropology of policy." "Cultural analysis", which the criticism against the "anthropology of policy" is tantamount to, intends to identify the meanings that individuals attach to development, and incorporate their viewpoints into development strategies on the assumption that everybody aspires to development. The aim of "cultural politics", on the other hand, is to deconstruct such "chronic realism." Instead it analyzes how the actual production of the meanings of development is implicated in the dynamics of everyday lives, in which the complex meshing of different agendas, arising out of daily social interactions, renders the identity of each individual a fluid, and incoherent entity. Therefore, development circulates in society in such a manner as to bring people under its sway, but does not permeate people's common sense. Similarly, policymakers also hold unstable and inconsistent subject positions, and constantly reflect on, and readjust the unfolding of policy directives. Given this contingent nature of officials' subjectivities, there exists a fertile ground for policy clients to influence how a policy is translated into action, even when it is executed in such a manner as to suppress the general public.

To counteract the manipulative nature of policy interventions, therefore, proponents of the "anthropology of policy" turn to political activism among marginal actors, which Escobar terms "self-defined and self-directed socioeconomic experiments",[73] or which Ferguson similarly calls "political engagement in one's own society."[74] As stated at the outset of this section, the "disciplinary" power inherent in the policy process constrains individuals' thoughts and actions. At the same time, "disciplinary" power is not only oppressive but also productive, in that it provides a frame of reference with which different actors renegotiate policy directions. As pointed out by Foucault with reference to the "governmentalization" of society, therefore, "the history of government as the 'conduct of conduct' is interwoven with the history of 'dissenting counter-conducts.'"[75] Given such interweaving of despotic and emancipatory forces, a policy forces, a policy intervention potentially bends back on itself, to self-correct "disciplinary" power intrinsic to policymaking endeavors, This is contrary to the assumption, widely held among critics of the "anthropology of policy", about the need to reconstitute policy directives themselves to countervail exclusionary forces.

This study combines the "actor-oriented" approach and the "anthropology of policy", in making sense of the policy process of "participatory" river control in Nepal. As described already, one of the drawbacks of the former approach is the failure to dissect wider socio-cultural dynamics, which lies outside the scope of the "actor-oriented" analysis of localized definitions of policy. The "anthropology of policy" makes up for the limitation by shedding light on the "governmentalization" of society that relegates the "participatory" river control policy to the imposition of pre-packaged flood mitigation measures. It thus highlights how the policy sets the overarching boundaries within which different actors contest the meanings of the policy. Another weakness of the "actor-oriented" approach, namely the "localism" of disregarding the potentiality of local people to influence the high-level policymaking process, can also be compensated for by the "anthropology of policy." This study analyzes the "cultural politics" surrounding the "participatory" river control policy, that is the contestation over its meanings among different players, varying from national policymakers to grassroots policy clients, whose subjectivities are multiple, fluid, and inconsistent. The intricate, dynamic interplay of unstable, incoherent life worlds of stakeholders at different levels, as described above, renders policymaking a non-linear, contingent process.

The final section of this chapter describes discourse analysis, with a view to elucidating the multiple, precarious nature of social actors' subjectivities. Though discourses offer a system of interpreting reality, thus serving to fix people's identities, they accords only partially consisted subjectivities, given their vulnerability to other competing and equally plausible counter-discourses. Before presenting discourse analysis, the following section critically reviews some literature on participatory development in the field of development studies, and considers how best to make sense of popular involvement in the policy process.

## "Citizen" Participation

This section turns to some studies on participatory development, with a view to critiquing how its proponents analyze public engagement with the policy process, drawing on the assessment of existing policy studies, undertaken in the preceding two sections. As stated in chapter 1, there exist increasing calls among proponents of participatory development to do away with the conventional "user group" focus in favor of broader people's involvement in the policymaking affecting their lives.[76] This parallels the rise of the "good governance" agenda[77] and the "rights-based" approach,[78] both of which point to the importance of public participation in policymaking, to hold government institutions accountable to the public, and to ensure that ordinary people fully exercise their "citizenship." In this way, people's engagement with the policy process has emerged as a converging concern of several thematic areas of development studies.

The "tyranny" critique of participatory development bears a similarity to the "anthropology of policy" in the sense that it stresses the perfidiousness of policymaking, by drawing on the Foucauldian notion of "disciplinary" power. According to this strand of studies, development programs aimed at activating popular participation pressurize local stakeholders into making a pretence of acting for the "greater good" of society, thereby catering to the priorities of external agents, rather than unleashing their self-organizing potential for societal change.[79] Proponents of participatory intervention are usually unaware of this insidious control, and resort to "methodological revisionism", only to depoliticize what should be a political process of addressing and countering the "disciplinary" forces inherent in planned participatory development.[80]

This "tyranny" critique has led to a group of research, which I would tentatively call "post-tyranny" studies, pointing out that the "tyranny" argumentation overgeneralizes the manipulative nature of participatory programs. The "post-tyranny" debate instead stresses that participatory interventions may turn out to be emancipative, depending on how local power struggles unfold. Participatory programs continue to be contested and reshaped in local contexts because "target groups" do not passively accept external interventions, but instead exercise popular agency[81] to rework them to suit location-specific social, political, and cultural particularities. We should avoid simplistically presuming that the "disciplinary" forces entailed in participatory programs are oppressive, but should instead look into the "micro-politics of participation as a situated practice."[82] This "post-tyranny" argument resonates with the "actor-oriented" approach, taken up above, according to which the execution of a policy directive provides an arena for "target groups" to contest power.

At the same time, the "post-tyranny" argumentation shares a weakness with the "actor-oriented" approach, that is its "localism" disregards the potentiality of ordinary people to influence overarching policymaking. Those associated with the "post-tyranny" debate tend to presuppose the necessity of a distinct set of initiatives to redress the tyrannical forces inherent in the policy process, with a view to "emancipating" the general public. In order to "enable" them to participate in high-level policymaking, it is necessarily to turn the beneficiaries into fully-fledged "citizens" capable of engaging in the policy process.[83] In their views, "citizen" participation in overarching policymaking, catalyzing transformation in broader socio-political circumstances, is diametrically opposed to the conventional mode of "user group" participation, facilitating people to take on the execution of externally supported programs. Such a polarized view is illustrated by the titles of some publications, such as "From Users and Choosers to Makers and Shapers" and "Making Rights Real."[84]

Such dichotomization is flawed in that a participatory intervention does not take place in a bounded, clearly demarcated space, but its impact spreads to a multitude of interrelating and overlapping sites.[85] This is because people's day-to-day social networks extend beyond the "project area", given their kinship and marriage ties, which extend beyond immediate neighborhoods, their economic activities connected with broader local, national, and even international economies, and their memberships in wider political communities. Moreover, drawing on their extensive networks, the general public react to, and resist decisions taken at a far distance, leading to an ongoing process of some form of political engagement with decision-making.[86] What further substantiates popular participation in the policy process is the "cultural politics" described in the preceding section. A policy directive is translated into a myriad of localized practices, which prompts policymakers to constantly review and readjust its overall direction. Given the unstable nature of policy elites' subjectivities, policymaking is inherently a contingent and self-reflexive process that provides a fertile ground for the general public to exercise leverage over its unfolding beyond their immediate localities.

Analysts of people's participation should therefore avoid immersing themselves in analyzing immediate, local-level outcomes of external interventions, with disregard to long-term, far-flung impacts. More importantly, in contrast to the "post-tyranny" argumentation that "citizen" participation needs be activated through external programs, the general public are already "citizens" capable of influencing high-level policymaking. The starting point should be the assumption that even a participatory intervention in a particular locality potentially offers space for its beneficiaries to influence not only how it is executed in the "project site", but also the overarching policy direction that has brought about the project. To build on Giddens' conception of "co-presence",[87] although grassroots beneficiaries and central policymakers may not be in direct contact, the former respond to and resist the latter's policies, which contributes to re-shaping the routinized manner in which policy interventions are designed and executed.

The "post-tyranny" advocacy of active "citizen" engagement with the policy process, however well-intentioned it may be, regulates the subjectivities of the general public, by insidiously compelling them to conform to the fallacy that they cannot exercise their popular agency without external interventions. "Citizens", in this case, are not the antithesis of "subjects", but are concurrently "subjects" who are subjugated to power and authority, whom Cruikshank succinctly describes as "citizen-subjects" as opposed to "citizen/subjects."[88] The "post-tyranny" strategy acts on the general public to act in their own "interests" fabricated by external agents, and is therefore tantamount to "technologies of citizenship" that facilitate outside agencies to pursue their agenda of promoting participation. In this regard, there exists a commonality between the "post-tyranny" approach "post-tyranny" approach and the more conventional mode of "user group" participation, in that both of them lapse into an instrument for external agents to get their beneficiaries to accept externally conceived programs.

# Nuanced Analysis of Discursive Practices

This chapter concludes by providing an analytical framework for the discursive practices of the actors involved in "participatory" river control at the national, district, and village levels. This final section begins by describing basic assumptions underlying the discourse analyses undertaken in this study, namely the "open texturedness" of discourse and the consequent "undecidability" of social identity. In this research, a discursive field is seen as an unstable ensemble of heterogeneous elements, drawn from a stock of multiple discourses. It does not confer a fixed, coherent identity on social agents, who are thereby required to take decisions by suturing rifts in their subjectivities. At the same time, discourse analysts should heed, not only the "undecidability" of discursive structures and subjectivities, but also enduring practices and processes. I consequently take up the notion of a "story-line" which exerts a strong influence with its discursive simplicity, and facilitates social actors, with divergent perceptions, to coalesce around and to communicate with one another. The final part of this chapter explains the Foucauldian perception of a "repressive hypothesis" which often serves as an underlying assumption of a "story-line", with its binary logic that simplistically categorizes social actors into "goodies" and "baddies." Discourse analysts can draw an important lesson from Foucault's problematization of a "repressive hypothesis."

## "Undecidability" of Discursive Structures

In this book, discourse analysis is concerned not only with texts and conversations, as has been the case with conventional linguistics, but also with how language constructs and enforces "appropriate and legitimate ways of practising development as well as speaking and thinking about it."[89] To use the definition in Hajer's work on environmental discourses, this study regards discourse as "a specific ensemble of ideas, concepts, and categorizations . . . through which meaning is given to physical and social realities."[90] In this sense, it is imperative both to pay attention to linguistic content, and also to analyze in which context a set of words and phrases is produced and mobilized, so as to understand its social backgrounds and effects.

    Underlying this approach are the assumptions that all objects and actions are meaningful, that their meanings are conferred by discourses articulating and contesting particular ways of interpreting social reality, and that discourses are constantly remolded by social actors' articulatory practices. One's articulatory practices can only partially fix meanings because a discourse cannot impose order with finality, but is penetrated by competing forces. The identity of a discourse is attained by distancing itself from counter-discourses that are equally plausible and would potentially reveal its contingency. A discourse is therefore dependent on, and at the same time, vulnerable to other discourses that are excluded from its discursive formation. Such "undecidability" of discursive structures brings about the "surplus of meaning",[91] or the "constant overflowing of every discourse by the infinitude of the field of discursivity."[92]

Discourses serve to fix the identities of objects and subjects in that they offer a system of understanding reality, and provide positions that social agents can identify with. At the same time, given the ultimate contingency of discourses, described in the preceding paragraph, social agents cannot attain fully constituted identities, but instead experience "social antagonism."[93] This study follows the position of Laclau and Mouffe that the central task of discourse analysts is to examine the "construction and contingent resolution of antagonistic relations",[94] namely how the identities of social agents are contested by other forces that stand at the limit of their order, and how social actors strive to fix their identities.

It is because of the contingency of discursive structures that social agents are forced to take decisions and act. In this respect, it is useful to distinguish, as pointed out by Laclau,[95] between "subject positions" and "political subjectivities." As opposed to conventional political analysis in which individuals are usually presumed to have given identities and interests, discourse analysis pays attention to shifting and multifaceted "subject positions" that social agents hold, owing to the inability of discourses to confer immutable, coherent identities.[96] On the other hand, individuals' "political subjectivities" are produced when they attempt to suture rifts in their social identities and to redefine their positions through the re-articulation of a particular discourse. Through such articulatory practices, different elements available in a discursive field are mobilized with an attempt to construct a seemingly coherent story. However, this process of "articulation" takes place only in parallel with "dislocation" that makes visible the inability of discursive structures to accord social agents stable positions. Social actors are thereby caught in the constant dilemma of having to, but of not being able to, attain consistent social identities.

## "Discourse Coalition" and "Story-line"

As pointed out by Fairclough,[97] it is crucial in discourse analysis to avoid the pitfall of overemphasizing the "social determination of discourse" or the "construction of the social in discourse." An advantage of focusing on such "decentering" of discursive structures as described in the preceding section is that it enables us to elucidate the dialectics between human agency and discursive structure. Rather than simplistically assuming that people's subjectivities are determined by, or constitute discursive structures, the approach drawing on La-clau and Mouffe stresses the constant interplay of the contingency of discourses and the necessity of agents to take decisions in the face of "social antagonism." At the same time, as pointed out by Howarth,[98] one major source of criticism is that it exaggerates the "undecidability" of structures and "subject positions", while downplaying more enduring practices and processes, such as ideological tradi-tions and organizational infrastructures. Laclau consequently acknowledges that, in order for political agents and movements to emerge, persisting discourses need be "available" and "credible."[99]

In reconciling the two positions, that is the one emphasizing the contingency of discursive structures, and the other elucidating persisting forces in discursive fields, Foucault states that "[w]henever one can describe . . . such a system of dispersion . . . one can define a regularity."[100] It is therefore imperative not only to heed the unstable and often contradictory nature of discursive structures, but also to take note of possible "regularity in dispersion." For this purpose, it is useful to mobilize Hajer's notions of "discourse coalition" and "story-line."[101] The two concepts of "discourse coalition" and "story-line" are founded on the assumption that discourse is not a coherent whole, but is continually reshaped through con-stant interactions and struggles among various actors with diverse articulatory practices. At the same time, because different actors interpret social reality in a myriad of ways, there needs be a platform for them to dissolve their differences, and to communicate with each other. Thus, a "story-line" serves as a common ground that allows different individuals and groups to form a "discourse coali-tion."

A "discourse coalition" differs from a traditional political alliance, in that various actors are considered to converge on a "story-line", rather than on "in-terests." Those involved in a coalition do not necessarily pursue a commonly shared strategy, but may even interpret "story-lines" differently, and therefore seek divergent commitments. In line with the post-structuralist notion of the provisional, non-static nature of "subject positions", referred to in the preceding section, Hajer considers actors' identities to be in the process of being reshaped constantly through the discursive practices that they engage in, to renegotiate their interpretations of reality. Discursive interactions are not merely a function of actors' a priori "interests", but also operate on, and influence their "subject posi-tions." "Discourse coalitions" are therefore not a stable and monolithic platform for actors with an enduring set of common "interests", but rather a fragmentary and fluid entity where dynamic interactions take place among actors with unsta-ble, diverse, and even contradictory "subject positions."

To exert a strong influence on social actors' perceptions, on the other hand, a "story-line" reduces problems of a complex nature into issues manageable within the existing social order. With its discursive simplicity and its resultant appeal, a "story-line" becomes established as taken-for-granted knowledge to which different actors assimilate themselves but do not entirely adhere, and to which they attach a myriad of meanings. This resonates with the assertion by Foucault that a "grid of interpretation"[102] is composed of both discursive and non-discursive elements, and that the latter denotes an "extra-discursive" system of domination. A "story-line" retains features of the latter in that it is established as a "regularity" in one's discursive field, and emanates "disciplinary" power that constrains one's articulatory practices in such a manner as to make them conform to social norms. Such an "extra-discursive" system of domination is often founded on a "repressive hypothesis", described in the ensuing section.

## "Repressive Hypothesis"

The Foucauldian term, "repressive hypothesis", was originally coined to designate an assumption underlying a "story-line" that permeates modern Western thinking concerning the relationships between knowledge, power, sexuality, and the body.[103] According to Foucault, the growing silencing of discourses and practices of sexual activities in the modern era, as opposed to the openness about sexuality in the classic period, is often attributed to the rise of capitalist production. It is usually assumed in Western societies that the bourgeoisie regarded sexual pleasures as counterproductive to capitalist production, and thus branded them as "perversions." Sexual discourses and practices are seen to stand in opposition to dominant societal "interests." As pointed out by Foucault, this "story-line", founded on a "repressive hypothesis" that dominant actors seek to suppress the general public, bolsters the notion of power prevailing in Western thinking, according to which power constrains freedom through prohibitions. According to the conventional "power-as-domination" view, moreover, the suppression of freedom needs be corrected through the dissemination of knowledge among ordinary people who can thereby challenge social domination.

The concept of a "repressive hypothesis" illuminates how the oppressive notion of power provides only a one-sided view. By drawing on a more productive conception of power, Foucault offers a refreshing synthesis of the growing suppression of sexual discourses and practices, which he opines did not necessarily emanate from the intentionality of the bourgeoisie per se. It was facilitated by more profound and far-reaching forms of power that organize, order, and regulate human bodies, that is the more productive facets of power downplayed by the above mainstream "story-line." As already described earlier, Foucault does not polarize power and knowledge, but regards them to be intrinsically related with each other in that the former upholds the latter, and facilitates individuals to discipline themselves to comply with, to respond to, or to resist societal norms. In parallel with the emergence of capitalist production was the increasing concern about relations between wealth, production, and populations, which has brought the survival and welfare of human species into the realm of governmental sur-

veillance. The rise of "bio-politics", which keeps track of various traits of biological existence, such as birth rate, longevity, and public health, disciplines
human bodies to self-monitor and self-measure their compliance with those
standards that are conducive to human welfare as well as to capitalist production.
The repression of sexuality can be attributed to this permeation of "bio-politics."

Foucault's problematization of a "repressive hypothesis" that elites seek to
suppress ordinary people points to the importance of refraining from analyzing
discursive interplays in terms of dyadic relationships between "goodies" and
"baddies", given the intermingling of "tyrannical" and "transformative" elements
in power dynamics. Domination and emancipation do not necessarily stand in
contradiction with each other, as is evidenced by how the suppression of sexuality
has taken place in tandem with the growing concern about human survival and
welfare. Contrary to such binary opposition as is assumed under the above
"story-line" on the repression of sexuality, therefore, emancipation is immanent
in social domination. This leads us back to the post-structuralist notion of the
provisional, multiple and contradictory nature of "subject positions", described
above. Dominant actors do not always seek to exercise control, but also facilitate
the emancipation of less powerful actors. Moreover, less dominant actors are not
passively subjugated to domination, but also assimilate themselves to productive
aspects of social control. As Foucault points out, "we must not imagine a world of
discourse divided . . . between the dominant one and the dominated one ... but as
a multiplicity of discursive elements that can come into play in various strategies."[104]

## Notes

    1. Peter Digester, "The Fourth Face of Power," *The Journal of Politics* 54, no. 4
(1992): 977-1007; Clarissa R. Hayward, "De-Facing Power," *Polity* 31, no. 1 (1998): 1-22.
    2. Robert A. Dahl, *Who Governs?: Democracy and Power in an American City* (New
Haven: Yale University Press, 1961).
    3. Peter Bachrach and Morton S. Baratz. *Power and Poverty: Theory and Practice*
(New York: Oxford University Press, 1970).
    4. Steven Lukes, *Power: A Radical View* (London: Macmillan, 1974).
    5. Steven Lukes, *Power: A Radical View*. 2d ed. (Basingstoke: Palgrave Macmillan
2005).
    6. As pointed out by Lukes (2005), the "power as domination" view did not exhaust
the entire terrain of debates on power in the 1960s, 1970s and 1980s, although the "power
as domination" view lay at the center of the debates among political theorists in those days.
Other prominent debates, referred to in Lukes' book (2005) included "beneficent power"
and "social power." For instance, "power-over" exists not only when A harms B's interests
but also when s/he satisfies and advances them. The latter type of "beneficent power"
manifests itself, for example, in parenting and apprenticeship. Moreover, A's power
emanates not only from human agency, but also from "social power" that s/he acquires by
being involved in institutionalized relations. The latter is exemplified by a judge who can

impose a death sentence. Such social actors' power is best described as a capacity to effect outcomes, as opposed to the more restricted "power as domination" view that regards power as the exercise or the vehicle of that capacity.

7. Joanne P. Sharp, Paul Routledge, Chris Philo and Ronan Paddison, "Entanglements of Power: Geographies of Domination/Resistance," in *Entanglements of Power: Geographies of Domination/Resistance*, eds. Joanne P. Sharp, Routledge Paul, Chris Philo and Ronan Paddison (London: Routledge, 2000), 1-42.

8. As pointed out by Foucault (1976), "disciplinary" power emerged around the eighteenth century, in an increasing number of settings, such as prisons, schools, hospitals, and workshops. In those arenas, ever-growing attention was paid to the idiosyncrasies of individual cases, to categorize human beings to measure their individual "quality" as healthy, hardworking, capable, law-abiding citizens. What added to this transformative process was the emergence of various scientific forms of classifications. A major feature of modern societies is the prevalence of "disciplinary" mechanisms that serve to shape, control, and monitor the behaviors of individuals.

9. Michel Foucault, "Afterword: The Subject and Power," in *Michel Foucault: Beyond Structuralism and Hermeneutics*, ed. Hubert L. Dreyfus and Paul Rabinow (Brighton: The Harvester Press, 1982), 213.

10. Michel Foucault, "Two Lectures," in *Power/Knowledge: Selected Interviews and Other Writings 1972-1977 Michel Foucault*, ed. Colin Gordon (New York: Harvester Wheatsheaf, 1980), 98.

11. Maarten A. Hajer, *The Politics of Environmental Discourse: Ecological Modernization and the Policy Process* (Oxford: Clarendon Press, 1995), 48-51.

12. David C. Hoy, "Power, Repression, Progress: Foucault, Lukes, and the Frankfurt School," in *Foucault: A Critical Reader*, eds. David C. Hoy (Oxford: Basil Blackwell, 1986), 142.

13. Foucault "Afterword," 220.

14. Foucault "Afterword," 211.

15. Foucault, "Two Lectures," 98-99. According to Foucault (1980, 99-101), for example, it can be deduced that, at the time of the industrial revolution in Europe, the bourgeoisie repressed "lunatics", whom they considered useless for capitalist production. Such a "descending" analysis would deadlock analysts, because it should have been in the interests of the bourgeoisie to provide skills training for the "insane" to become productive forces, rather than excluding them. On the other hand, an "ascending" analysis brings to light that the exclusion of "lunatics" did not derive from the intentionality of the bourgeoisie per se. On-the-ground investigations reveal that "disciplinary" forces had already been at work throughout society to marginalize those who did not meet the standards of "sanity", and that the bourgeoisie made use of this mechanism to uphold the value of industrialization. Foucault demonstrates, drawing on this case, the importance of grasping what is happening at the micro-level, before revealing a more general picture of power dynamics.

16. Foucault "Afterword," 222.

17. Michel Foucault, *Discipline and Punish: The Birth of the Prison* (London: Penguin Books, 1976), 25.

18. Michel Foucault, "Questions of Methods," in *The Foucault Effect: Studies in Governmentality*, eds. Graham Burchell, Colin Gordon and Peter Miller (Chicago: The University of Chicago Press, 1991a), 75.

19. Digeser, "The Fourth Face," 991.

20. Digeser, "The Fourth Face," 1004.

21. Philip Barker, *Michel Foucault: An Introduction* (Edinburgh: Edinburgh University Press, 1998), 38-39.

22. Digester, "The Fourth Face," 991.

23. Foucault, "Two Lectures," 98.

24. Anthony Giddens, *Constitution of Society* (Cambridge: Policy Press, 1984).

25. Stewart R. Clegg, *Frameworks of Power* (London: Sage Publications, 1989), 101-3.

26. Lukes, *Power* 2d ed., 56.

27. As stated in the main text, Giddens considers that structure, as both the medium and the outcome of human actions, is not external but closely tied to human agency. According to Giddens (1984, 16-17), routinized social practices exhibit dynamic "structural properties" that are constantly produced and reproduced by human actions, rather than a static "structure" that one-sidedly constrains the freedom of social actors.

28. Giddens, *Constitution of Society*, 3.

29. Bourdieu's concept (1977) originates from a similar concern to resolve the structure/agency antinomy, and brings to light how human actions are determined by the nexus between structure and agency. According to Bourdieu, social structure constitutes constraints for human thoughts and actions, which are also internalized by social agents as a practical understanding of the boundaries of what is possible. This "duality of structure" enables individuals to acquire a "feel for the game" in line with objective probabilities of success or failure. "Habitus" thus acquired is "the system of structured, structuring dispositions", to use the phrase found in Bourdieu's work (1990, 52) which sets structural limits on, as well as generates, social actions. In this way, just as with Giddens' "structuration" view, "habitus" serves to transcend the structure/agency dichotomy that plagues the conventional "power as domination" view.

30. James C. Scott, *Domination and the Arts of Resistance: Hidden Transcripts* (New Haven: Yale University Press, 1990).

31. Anthony Giddens, *Profiles and Critiques in Social Theory* (London: Macmillan, 1982), 4.

32. Anthony Giddens, *Central Problems in Social Theory: Action, Structure and Contradiction in Social Theory* (London: Macmillan, 1979), 91.

33. Giddens, *Central Problems*, 91.

34. Scott, *Domination and the Arts*, 93.

35. Giddens, *Constitution of Society*, 15.

36. Barker, *Michel Foucault*, 38-39.

37. Wayne Parsons, *Public Policy: An Introduction to the Theory and Practice of Policy Analysis* (Aldershot: Edward Elgar, 1995), 151-53.

38. Geof Wood, ed., *Labelling in Development Policy* (London: Sage, 1985).

39. Donald A. Schon and Martin Rein. *Frame Reflection: Toward the Resolution of Intractable Policy Controversies* (New York: Basic Books, 1994).

40. Michael Hill, *The Policy Process in the Modern State*. 3d ed. (Hertfordshire: Prentice Hall/Harvester Wheatsheaf, 1997), 63.

41. Alberto Arce, Magdalena Villarreal and Pieter De Vries, "The Social Construction of Rural Development: Discourses, Practices and Power," in *Rethinking Social Devel-*

*opment: Theory, Research and Practice*, ed. David Booth (Essex: Longman, 1994), 153-9; Norman Long, "From Paradigm Lost to Paradigm Regained," in *Battlefields of Knowledge: The Interlocking of Theory and Practice in Social Research and Development*, eds. Norman Long and Ann Long (London: Routledge, 1992), 34-35; Norman Long, *Development Sociology: Actor Perspectives* (London: Routledge, 2001), 30-48.

42. Long, *Development Sociology*, 25.

43. Long, "From Paradigm Lost," 22.

44. Several studies on natural resource management, such as those by Li (1996) and Mosse (1997), also confirm that development interventions are best understood by investigating such "interfaces." These studies illustrate that interventions provide both material and symbolic resources, for which various social actors contest, in order to enunciate, or challenge existing patterns of domination and control.

45. Alberto Arce, "Re-Approaching Social Development: A Field of Action between Social Life and Policy Process," *Journal of International Development* 15 (2003): 848.

46. Paul A. Sabatier and Hank C. Jenkins-Smith, *Policy Change and Learning: An Advocacy Coalition Approach* (Boulder: Westview Press, 1993); Martin J. Smith, *Pressure, Power and Policy: State Autonomy and Policy Networks in Briton and the United States* (New York: Harvester Wheatsheaf, 1993). ′

47. Edward J. Clay and Bernard B. Schaffer, *Room for Manoeuvre: An Exploration of Public Policy in Agricultural and Rural Development* (London: Heinemann Educational Books, 1984); Merilee S. Grindle and John W. Thomas, *Public Choices and Policy Change: The Political Economy of Reform in Developing Countries* (Baltimore: Johns Hopkins University Press, 1991).

48. Michael Lipsky, *Street-Level Bureaucracy: Dilemmas of the Individual in Public Services* (New York: Russell Sage Foundation, 1980).

49. Arce, Villarreal and De Vries, "The Social Construction," 152-71; Norman Long ,"Exploring Local/Global Tranformations: A View from Anthropology," in *Anthropology, Development and Modernities: Exploring Discourses, Counter-Tendencies and Violence*, eds. Alberto Arce and Norman Long (London: Routledge, 2000), 184-201.

50. Long "Exploring Local/Global", 192.

51. Paul Hebinck, Janden Ouden and Gerald Verschoor, "Past, Present, and Future: Long's Actor-Oriented Aproach at the Interface," in *Resonances and Dissonances in Development: Actors, Networks and Cultural Reportoires*, eds. Paul Hebinck and Gerald Verschoor (Assen: Royal Van Gorcum, 2001), 11.

52. Clegg, *Frameworks*, 147-48.

53. Alberto Arce, "Experiencing the Modern World: Individuality, Planning and the State," in *Resonances and Dissonances in Development: Actors, Networks and Cultural Reportoires*, eds. Paul Hebinck and Gerald Verschoor (Assen: Royal Van Gorcum, 2001), 104-5.

54. Alberto Arce, "Experiencing the Modern," 105.

55. Quoted in Hebinck, Ouden and Verschoor, "Past, Present", 11.

56. Long, *Development Sociology*, 53.

57. Long, "Exploring Local/Global", 198-200.

58. Long, *Development Sociology*, 70.

59. Long, *Development Sociology*, 53.

60. Cris Shore and Susan Wright, eds., *Anthropology of Policy: Critical Perspectives on Governance and Power* (London: Routledge, 1997); Raymond Apthorpe and Des

Gasper, eds., *Arguing Development Policy: Frames and Discourses* (London: Frank Cass, 1996).

61. Michel Foucault, "Governmentality," in *The Foucault Effect: Studies in Governmentality*, eds. Graham Burchell, Colin Gordon and Peter Miller (Chicago: The University of Chicago Press, 1991b), 87-104.

62. Foucault, *Discipline and Punish*, 221.

63. Foucault, "Governmentality," 102.

64. Shore and Wright, *Anthropology of Policy*.

65. Hubert L. Dreyfus and Paul Rabinow, *Michel Foucault: Beyond Structuralism and Hermeneutics* (Brighton: The Harvester Press, 1982), 196.

66. Bernard B. Schaffer, "Towards Responsibility: Public Policy in Concept and Practice," in *Room for Manoeuvre: An Exploration of Public Policy in Agricultural and Rural Development*, eds. Edward J. Clay and Bernard B. Schaffer (London: Heinemann Educational Books, 1984), 146.

67. Escobar (1995, 102-53), for example, highlights how development policies usually construct a causality of malnutrition, as if it were uninfluenced by location-specific socio-political situations, so as to legitimize the straightjacket application of standardized integrated development programs. Ferguson (1990) analyzes the manner in which Lesotho's development policy depicts the country as a primitive agricultural economy in need of being modernized through the provision of technologies and infrastructure. Pigg (1992) points out that development policies in Nepal implicitly oppose the "village" and "progress", and that the notion of "development" is assimilated into the ways the "villagers" see themselves and relate to others, to bring about the assumption prevalent among the "villagers" that "development" is brought in from outside. Crewe and Harrison (1998) explain how the policies of some aid agencies operating in Sri Lanka and Zambia embody some flawed assumptions, such as the notions of development as a linear progression, and technologies as morally neutral, thereby legitimatizing external interventions to bring in exogenous know-how and resources.

68. For example, see Gasper (1996), Gardner and Lewis (1996), Kiely (1999), Storey (2000), and Naderveen Pieterse (2001).

69. Arturo Escobar, "Beyond the Search for a Paradigm: Post-Development and Beyond," *Development* 43, no. 4 (2000): 11-14.

70. Escobar, "Beyond the Search," 14. Emphasis added by this author

71. Donald Moore, "The Crucible of Cultural Politics: Reworking 'Development' in Zimbabwe's Eastern Highlands," *American Ethnologist* 26, no. 3 (2000): 673.

72. Honor Fagan, "Cultural Politics and (Post) Development Paradigm(S)," in *Critical Development Theory: Contributions to a New Paradigm*, eds. Robaldo Munck and Denis O'Hearn (London: Zed Books, 1999), 178-95.

73. Arturo Escobar, *Encountering Development: The Making and Unmaking of the Third World* (Princeton: Princeton University Press, 1995), 223.

74. James Ferguson, *The Anti-Politics Machine: "Development," Depoliticization, and Bureaucratic Power in Lesotho* (Cambridge: Cambridge University Press, 1990), 287.

75. Michel Foucault, *Lecture* (Pasquino: College de France, 1978), cited in Colin Gordon, "Governmental Rationality: An Introduction," in *The Foucault Effect: Studies in Governmentality*, eds. Graham Burchell, Colin Gordon and Peter Miller (Chicago: The University of Chicago Press, 1991), 5.

76. John Gaventa and Camillo Valderrama, *Participation, Citizenship and Local Governance: Background Paper for Workshop, Strengthening Participation in Local Governance* (Brighton: IDS, 1999), 1-6; Andrea Cornwall and John Gaventa, *From Users and Choosers to Makers and Shapers: Repositioning Participation in Social Policy* (Brighton: IDS, 2001), 9-19.

77. Andrea Cornwall, *Beneficiary, Consumer, Citizen: Perspectives on Participation for Poverty Reduction* (Stockholm: SIDA, 2000), 61-62.

78. John Gaventa, "Introduction: Exploring Citizenship, Participation and Accountability," *IDS Bulletin* 33, no. 2 (2002): 3-6.

79. For example, see Mosse (2001), and Kapoor (2002).

80. Bill Cooke and Uma Kothari, "The Case for Participation as Tyranny," in *Participation: The New Tyranny?*, eds. Bill Cooke and Uma Kothari (London: Zed Books, 2001), 1-15.

81. Sam Hickey and Giles Mohan, "Towards Participation as Transformation: Critical Themes and Challenges," in *Participation, from Tyranny to Transformation?: Exploring New Approaches to Participation in Development*, eds. Sam Hickey and Giles Mohan (London: Zed Books, 2004a), 3.

82. Andrea Cornwall, *Making Spaces, Changing Places: Situating Participation in Development* (Brighton: IDS, 2002), iii.

83. For example, see Cornwall and Gaventa (2001), Gventa (2002), and Hickey and Mohan (2004b).

84. These are the titles of Cornwall and Gaventa (2001) and of the IDS Bulletin (33, no.2) which includes Gaventa's article (2002), respectively.

85. Roderick L. Stirrat, "The New Orthodoxy and Old Truths: Participation, Empowerment and Other Buzz Words," in *Assessing Participation: A Debate from South Asia*, edited by Sunil Bastian and Nicola Bastian (Delhi: Konark Publishers, 1996), 71-72; Hickey and Mohan, "Towards Participation," 16-18.

86. Glyn Williams, "Evaluating Participatory Development: Tyranny, Power and (Re)Politicisation," *Third World Quarterly* 25, no. 3 (2004): 557-78.

87. Social interactions take place across time and space, without regard to direct connections between those engaged in them. As pointed out by the "structuration" theory, described earlier, individuals' conducts are bounded by the structural elements of society, and thus exhibit a repetitive and routinized nature. At the same time, social actors do not simply pursue their routines, but also strive to adjust their daily flow of conduct in order to turn "structural properties" to advantage. By way of taking actions, social actors also contribute to the reproduction and remolding of "structural properties." Giddens (1984, 142) thereby advances a notion of "co-presence" that denotes such potentiality of human agency to exert influence "across tracts of spaces broader than those involved in immediate contexts of fact-to-face interactions."

88. Barbara Cruikshank, *The Will to Empower: Democratic Citizens and Other Subjects* (Ithaca: Cornell University Press, 1999), 24.

89. Ralph D. Grillo, "Discourse of Development: The View from Anthropology," in *Discourse of Development: Anthropological Perspectives*, eds. Ralph D. Grillo and Roderick. L. Stirrat (Oxford: Berg, 1997), 12.

90. Hajer, *Environmental Discourse*, 44.

91. Ernesto Laclau and Chantal Mouffe, *Hegemony and Socialist Strategy*. 2d ed. (London: Verso, 2001), 111.

92. Laclau and Mouffe, *Hegemony and Socialist*, 113.

93. Laclau and Mouffe, *Hegemony and Socialist*, 122.

94. David Howarth and Yannis Stavrakakis, "Introducing Discourse Theory and Political Analysis," in *Discourse Theory and Political Analysis: Identities, Hegemonies and Social Change*, eds. David Howarth, Aletta J. Norval and Yannis Stavrakakis (Manchester: Manchester University Press, 2000), 10.

95. Ernesto Laclau, *New Reflections on the Revolution of Our Time* (London: Verso, 1990), 60-61.

96. Howarth and Stavrakakis, "Introducing Discourse," 6.

97. Norman Fairclough, *Discourse and Social Change* (Cambridge: Polity Press, 1992), 65.

98. David Howarth, *Discourse* (Buckingham: Open University Press, 2000), 121-22.

99. Laclau, *New Reflections*, 65-67.

100. Michel Foucault, *The Archaeology of Knowledge* (London: Tavistock, 1986), 38.

101. Hajer, *Environmental Discourse*, 12-13.

102. Dreyfus and Rabinow, *Michel Foucault*, 121.

103. Michel Foucault, *The History of Sexuality, Vol.1: Introduction* (Harmondsworth: Penguin, 1979).

104. Foucault, *The History of Sexuality*, 100-1.

# Chapter 3
# The Rise and Fall of "Participatory" River Control at the Center

This chapter undertakes a detailed examination of the policy process of "participatory" river control, with a focus on key players at the national level, that is political leaders, and bureaucrats at the Department of Irrigation (DOI). The decentralization and participation (D&P[1]) agenda, which held sway in the discursive field of Nepal's planned development in the 1990s, spilt over to cause the "participatory" river control policy to emerge as an unobjectionable option, in the absence of specific deliberations. Accordingly, the constrictive nature of the D&P rhetoric, to reduce policy issues of a complex nature to an agenda amenable to planners' blueprints, was also implanted in the "participatory" river control policy. The first section of this chapter analyzes how the policy served as a "political technology", to mask possible alternative approaches to flood mitigation, other than the delivery of pre-packaged engineering solutions.

This chapter then looks into how the two main actors at the national level interpreted, and translated into action, the "participatory" river control policy. Politicians and the DOI bureaucracy coalesced around the conventional "story-line" of development, which treated villages as backwaters trapped in outmoded ways of living. As a result, they generally victimized the flood-affected populace and equated flood mitigation with the distribution of pre-packaged technical solutions. While the overall tone of "participatory" river control was set in this way, at the same time, central policymakers were caught in a dilemma over whether or not to relegate the policy to such a centrally directed process. The "mainstream" notion of local river control as the delivery of handouts was vulnerable to other counter-discourses that argued for more emancipative approaches that would accord local autonomy. Central policymakers therefore felt ambivalent about the blueprint embodied in the policy, and their discursive practices constituted an unstable ensemble of heterogeneous elements. The policymaking process of "participatory" river control therefore did not proceed in a linear and "logical" manner. On the contrary, the way "participatory" river control was translated into action was subjected to ceaseless renegotiations among central stakeholders.

Towards the end of the 1990s, the central government came to ignore the "participatory" river control policy, and to bypass the locally elected chair of the district governing body. It is possible to make sense of this policy decline in the light of the political circumstances in the late 1990s that inclined political leaders in the ruling party to drop the policy. Owing to the frequent discrepancies between the party in power at the center and that at the district level, the value of the policy declined as a means for central politicians to distribute patronage among their local party cadres. This "descending" analysis of this policy decline, albeit ostensibly reasoned and unquestionable, is biased toward a zero-sum notion of power that dichotomizes dominant and powerless actors, and sets them up in opposition as "baddies" and "goodies." As stated above, how central policy elites made sense of "participatory" river control was in a state of flux, wavering between their propensity to inhibit people's participation, and their need to do away with their exclusionary practices. This chapter therefore concludes by presenting an alternative "ascending" analysis of focusing on the unstable and inconsistent nature of the identities of policymakers.

# Emergence of "Participatory" River Control

The local-level river control portfolio of the government, which is the topic of this study, was not immune to the D&P agenda of the 1990s. This section describes how the policy on local river control was founded on an interpretative process of manufacturing "reality", within a particular framework in which flooding problems and solutions were constructed. As pointed out by the "argumentative approach" and the "anthropology of policy", explained in chapter 2, a policy objective is usually given the appearance of a foregone conclusion in the disguise of impartial language, thus diverting attention away from other possible alternatives, and inhibiting policy stakeholders from engaging in policy discussions. This was also the case with the local river control sector. The D&P rhetoric contained in the policy, with its symbolic function of evoking an impression of the government's commitment to public welfare, served to mask potential measures other than providing handouts to flood-affected areas. In line with the entire D&P process described at the beginning of this chapter, the age-old "story-line" of development exerted a strong influence local river control endeavors, in that local river control was relegated to the delivery of pre-packaged flood control projects.

    It is to be noted that, although the overall tone of local-level river control was driven by the "story-line", its policymaking was a highly contested process, even including forces that diametrically opposed distorting the D&P agenda. The ostensibly coherent stories provided in this section highlight only certain aspects of the policy process that elucidate the symbolic and manipulative nature of the policy. The main stakeholders at the center, namely political leaders, and bureaucrats at the Department of Irrigation (DOI), did not simply collude with one another to blatantly reduce local river control to centrally directed endeavors. On the contrary, the policy process was covertly but constantly renegotiated by these policy elites at the center whose identities were fluid in the face of counter-discourses that point to the precariousness of the way local river control was translated into action. The accounts provided below are therefore not intended to tended to provide the entirety of how policymakers made sense of local river control, but instead are a mere attempt at the "construction and contingent resolution of antagonistic relations", to use the phrase found in Howarth and Stavrakakis' work.[2]

## Context and Policy Objectives

In Nepal, as of May 2001, there were seventy-five districts which, within their jurisdictions, encompass fifty-eight towns and 3912 villages. These administrative units were headed by elected representatives who formed the executive committees, namely the District Development Committee (DDC), the Village Development Committee (VDC), and the Municipality (see Figure 3.1). The VDC and the Municipality were further divided into nine or more electoral districts, in each of which a Ward Committee was formed though direct election. The Ward Committee was composed of one chair and four members.

**Figure 3.1. Institutional Set-up**

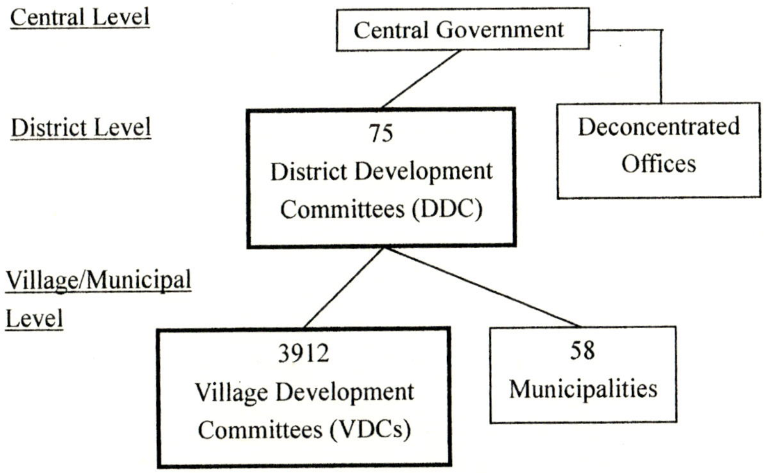

Central Level — Central Government

District Level — 75 District Development Committees (DDC) — Deconcentrated Offices

Village/Municipal Level — 3912 Village Development Committees (VDCs) — 58 Municipalities

Nepal is divided into three physiographic zones, the mountains, the hills, and the Tarai plain (see Map 1.1). The DOI was responsible for flood control in the Tarai until 2002. River control activities of the DOI were classified into national and district-level projects. The DOI's central office directly operated national-level undertakings, which resulted in the construction of continuous dikes along stretches of rivers. On the other hand, district-level river control projects, the subject of this study, were implemented through the District Irrigation Office

(DIO), that is the deconcentrated arm of the DOI. The latter group of projects usually protected a limited number of settlements, through the construction of small-scale structures, such as groins or spurs to reduce water velocities and divert flows away from riverbanks (see cover photo), and retaining walls to protect riverbanks.

In line with the growing critical reflections among Nepal's policymakers that top-down, supply-driven approaches rarely worked to accelerate rural development, the DOI also started placing emphasis, in the late 1980s, on farmers' involvement in the construction and management of irrigation systems.[3] This was also the case with the Department's river control portfolio, which began to be executed through the DIO, in the 1980s, as per the 1982 Decentralization Act that had mandated central line agencies, including the DOI, to classify their undertakings into central and district-level activities. However, the DIO, which was set up ostensibly to open up the Department's hierarchy to public scrutiny, implemented river control projects, with little consultation with local leaders and beneficiaries.

The situation changed after a landmark policy directive was issued in 1992 by the then Minister of Water Resources. First, though the DOI was to retain central control over the allocation of resources to different flood-affected districts, the policy directive stipulated that where and how to use the allotment would be left to the DDC chair who would form the Implementation Committee, by drawing members from other local politicians, usually DDC members (see Figure 3.2). The DIO was to serve as the secretary to the committee, and would be compelled to implement its river control portfolio according to decisions made by the

**Figure 3.2. Institutional Route of Policy Flow:**
**"Participatory" River Control**

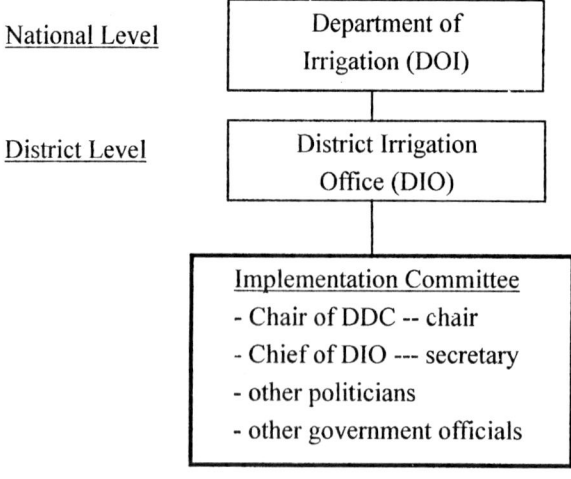

committee. If and when necessary, the committee was to draw its members from other government offices. Second, the policy recommended that local river control projects should be implemented with the participation of beneficiaries. The government had been providing two types of resources for district-level river control projects, namely materials such as galvanized iron (G.I.) wire and sandbags, as well as financial resources. When the above policy was followed, the financial resources were to be utilized to procure additional materials, such as boulders and sandbags, instead of being spent on contractors, and local residents were expected to contribute unpaid labor for the construction of river control structures.

In fact, this policy directive, hereinafter called the "participatory" river control policy, was originally intended to cover one particular project in eastern Nepal (see Appendix). The instruction eventually came to be treated as a nationwide guideline, in line with the decentralization and participation (D&P) agenda that held sway in Nepal's planned development after the 1990 political change. The *Panchayat* system, in which the king assumed sovereignty as a symbol of the national unity, ended in 1990 when various political parties joined hands in organizing mass anti-*Panchayat* movements, building on a growing discontent among the general populace with the regime beset with economic stagnation and a series of corruption scandals. A new constitution was enacted subsequently to espouse a multi-party system under a constitutional monarchy.[4] One of the driving forces behind the political movement leading to the collapse of the monarchical *Pahchayat* rule was the rising expectation among ordinary people that democratization would improve their living standards, and that the fruits of development monopolized by a small number of the privileged would be more equitably shared. As a result, the post-1990 governments officially treated the D&P agenda as an integral part of development programs.

The replication of the Minister's 1992 directive for "participatory" river control came about as a foregone, unobjectionable conclusion, not as a result of, but in default of particular deliberations. It is not plausible to pinpoint particular decisions or documents that signaled the initiation of the river control policy, given the absence of formal discussions as to whether to apply the Minister's order nationwide. As part of the D&P strategy, the central government had been forming various district-level sectoral committees, to bring local-level operations of central line agencies under the supervision of local political leaders. Moreover, "participation" had become an orthodox approach among central line agencies, especially among those concerned with natural resource management, such as forestry, soil conservation, and irrigation. The 1992 directive was promulgated even in the form of a letter as an administrative order. The Minister did not opt to go through a bureaucratic process of having it gazetted as a fully-fledged policy, which would have required prior consultations with civil servants.

## "Participatory" River Control as a "Political Technology"

In line with the "anthropology of policy", explained in chapter 2, the D&P agenda entailed an interpretative process of fabricating "reality", to covertly uphold the expansion of state interventions in society, or what Foucault terms the "govenrmentalization" of society. Such "disciplinary" power arose from a "story-line" of development, which had been prevailing in the history of Nepal's planned development, and had set the overall tone of the ostensibly "new" D&P strategy. The "story-line" polarizes the center and the periphery on the scale of progress and backwardness by producing the construct of "villagers" who are isolated, traditional and ignorant.[5] The resultant assumptions are that "development" exists outside "villages", that knowledge and resources need to be brought in from outside, and that the central government is the vehicle for social and economic modernization. A view widely held among researchers on Nepal[6] is that the post-1990 D&P reform was relegated to a centrally directed process of infusing external knowledge and resources into villages, instead of unleashing grassroots activism for social change.

The D&P agenda was also spilt over to the "participatory" river control policy in such a way as to relegate flood mitigation to the delivery of handouts to the affected populace, rather than promoting local-level initiatives. Flood mitigation measures are not limited to structures using iron wire and sandbags, which is not readily available in local areas, but also include approaches that encourage local coping strategies for the flood-affected populace to cope with flooding drawing on local resources. Vegetation along riversides can contain the force of flood-water and prevent bank erosion. There are also non-structural measures, including land use regulations, flood forecasting, warning and evacuation, and disaster insurance. In addition to these types of local coping strategies, it is widely believed in Nepal that upper watershed conservation can reduce water run-off and soil erosion, thereby containing flooding in downstream areas.[7] The "participatory" river control policy enforced closure on these other alternatives, and diverted attention away from these possible solutions.[8]

More importantly, a natural event would not turn into a disaster, unless people were compelled to reside or undertake livelihood activities in flood-prone areas, and lacked sufficient assets and reserves to withstand shocks. The degree to which individuals and groups are vulnerable to damage and suffering is often a function of the location of their residences and farmland, as well as the structures and types of housing.[9] In rural Nepal, generally, the disadvantaged are more likely to reside in shanties vulnerable to flooding, and farm on marginal flood plains.[10] Disaster reduction therefore has to address such socio-political dynamics in which the livelihoods of marginal groups are more likely to be put at risk, in addition to finding technical solutions to counteract the forces of nature. However, the "participatory" river control policy turned a blind eye to the social, economic, and political circumstances that rendered certain individuals or groups particularly vulnerable to floods.

Moreover, the country's southern neighbor, India, played a part in constricting the scope of Nepal's river control efforts. Although the construction of continuous dikes could have been a more effective solution in mitigating the damage in the Tarai plain, such efforts would inflict severer flooding on northern India, as a result of the increased channel capacity to carry floodwater across the border. The government of India had not only blocked Nepal's initiative to construct large-scale structures, but had even built barrages near the border along some of the major rivers, thereby causing floodwater to stagnate on the Nepalese side during the rainy season. Despite such interventions by its southern neighbor, Nepal had little option but to remain on good terms with India. Nepal is highly dependent on India for its trade and commerce, since it has few alternative land routes to the sea except those though Indian territories. Moreover, India historically exerted a strong influence over the domestic politics of Nepal, and it was difficult for the Nepalese government to manage its water resources without the consent of India.

The "participatory" river control policy was therefore tantamount to what Foucault terms a "political technology"[11] that reduces what is essentially a political issue to a technical matter. It not only ignored underlying socio-political forces that rendered residents in the Tarai vulnerable to floods, but also ruled out other approaches, such as local coping strategies and upper watershed conservation. "People's participation" upheld in the policy was therefore reduced to a procedural matter of getting people to participate in pre-packaged river control efforts using such materials as iron wire and sandbags, rather than allowing them to shape and seek their own strategies. As illustrated by one of its clauses that "(l)ocal beneficiary groups will be actively involved in the implementation" (see Appendix), the policy implicitly took for granted that the affected populace would be willing to provide their contributions to river control structures, the parameter of which was pre-determined by the government. In this way, the policy invoked an imagery that the government was responding to the "needs" and "wants" of the general public who needed to protect their villages through the construction of river control structures using iron wire and sandbags.

The actual unfolding of the policy, however, did not proceed in such a "logical" and purposive manner. On the contrary, the way the policy was translated into action was ceaselessly reflected on, and renegotiated among central-level policymakers who did not necessarily seek to equate "participatory" river control with the top-down process of distributing pre-packaged projects. As is inferred from the "undecidability" of discourse, explained in chapter 2, the discursive structure of the policy was vulnerable to other alternative definitions of "participatory" river control, such as those taken up in this section. This "open texturedness" of the river control policy facilitated "cultural politics" over the policy direction among different stakeholders, varying from policy elites to target groups. This is in line with the assertion of the "anthropology of policy" that "disciplinary" power inherent in policymaking is both oppressive and productive. Policymaking insidiously upholds the "governmentalization" of society, but at the same time, provides a frame of reference that social actors draw on to renegotiate the policy direction. The next section focuses on the "cultural politics" among national policymakers, while the ensuing three chapters deal with those among various local actors.

# "Regularity in Dispersion" of Policymakers' Stances

Underlying the seemingly reasoned accounts of the "participatory" river control, provided in the preceding section, was the "story-line", explained in the preceding section, that regards development as a vertical process of bringing external resources into outlying areas. Central-level actors concerned with local river control endeavors coalesced around the conventional "story-line" that patronizes people as passive recipients of development, thereby treating the flood-affected populace as victims and taking for granted the distribution of pre-packaged physical structures. This section illustrates how the "story-line" exerted a strong influence, and yet was constantly challenged within the discursive fields of the two major players in the local river control portfolio, that is political leaders affiliated with major political parties, and officials at the DOI. As stated below, the main actors felt ambivalent about the blueprint embodied in the policy, and their articulatory practices constituted an unstable ensemble of heterogeneous elements, including "heretical" ones diametrically opposed to the top-down nature of the policy.

## National Political Leaders

Flood mitigation can produce direct political payoffs, since stakes are high for those who are exposed to the risks of losing their property or source of livelihoods. In the eyes of political leaders, river control constituted an instrument for their local party cadre to strengthen their support bases. The focus of the D&P strategy in the river control sector therefore tended to be placed on patronage distribution, rather than on local autonomy. This is attested to by the occasional changes in the district-level institutional set-up for "participatory" river control projects. The central government made the modifications, in response to the discrepancy between political parties assuming power at the center and enjoying a majority at the district level.

Under the multi-party system in the 1990s, two major political parties competed for a majority in the legislature, that is the Nepali Congress (NC), and the Community Party of Nepal-Unified Marxist-Leninist (CPN-UML). The other major party was the *Rashtriya Prajantra* Party (RPP) that was formed by former *Panchayat* leaders after the 1990 political changes.[12] Although the major political parties professed their commitment to the D&P agenda in their party manifestos,[13] they regarded it largely as a means of strengthening their networks through resource allocation. This is symbolized by the alleged manipulation by the central government of the results of the local elections held under the NC government in 1992, and under the CPN-UML-led government in 1997. As a result, the NC occupied a majority of the DDC posts between 1992 and 1997, while the CPN-UML dominated the DDCs after 1997. Given the frequent changes of central governments,[14] at the same time, there were often discrepancies between the ruling party at the center and the dominant party at the district level, which often prompted the central government to bypass the DDC chair.

In the case of the river control sector, the CPN-UML government in 1995 instructed the DDC chair to make it a rule to consult with representatives of all major political parties, to counterbalance the NC that occupied about eighty percent of the posts for the DDC chair at the time. The Communist Government argued that the preceding Congress Government had allowed DDC chairs, who headed the district-level committee for river control, to use the sector's resources for partisan purposes, and defended the reconstitution of the Committee as a countermeasure against favoritism. Furthermore, in 2000, with a view to sidelining DDC chairs, an overwhelming majority of whom belonged to the CPN-UML, the NC Government designated the Chief District Officer (CDO), a government bureaucrat, to head the district-level committee.

The conventional "story-line" that treats development as exogenous to villages was useful in diverting attention, away from such political considerations underlying the "participatory" river control policy. Political leaders accordingly tended to describe the flood-affected populace as the passive victims of Nepal's unfavorable natural conditions. Moreover, the general public was usually depicted as too helpless and uneducated to care about the consequences of their reckless behaviors of encroaching on flood plains. The resultant approach should be to distribute technical packages, rather than attempting to upgrade people's capacity to anticipate, adjust to, and recover from the impacts of hazards. "Participation" was therefore equated by many political leaders with an instrument to have pre-packaged river control projects accepted by villagers, and to enhance the projects' cost-effectiveness by using people's unpaid labor. It was not aimed at enabling people to address broader socio-political issues that put people at risk.

A close examination of politicians' articulatory practices, at the same time, reveals that they were far from being monolithic and immutable. On the contrary, their discursive field encompassed "heretical" elements that even resented relegating river control to the technical matter of lessening the impact of natural hazards through engineering solutions. This "heterogeneity" derived from the "undecidability" of discursive structures, explained in chapter 2.[15] As described above in this chapter, flooding in the Tarai was exacerbated by various socio-political forces that rendered its residents vulnerable. The "participatory" river control policy, which ignored these socio-political issues, was inherently vulnerable to other interpretations that would disclose the negligence. As a result, there existed a fertile ground for various counter-discourses to penetrate into the discursive practices of politicians. Some political leaders accordingly considered that the river control policy detracted from the country's D&P strategy that supposedly delegated decision-making power to local people. For example, a former Minister of Water Resources reflected on the way the river control portfolio had been managed in the 1990s, and pointed out the need to do away with perfunctory "participation."

> "I think river control projects should come from below, basically. And the resources should be given to the district as the river control fund. Then, you know, local people will decide how it is to be used. The central government should give general guidelines. Ten percent, twenty percent [of the costs] should be covered by popular participation, the maintenance should be in the hands of the people, and so on. Everything else, all other decisions, should be left to the people."[16]

At the same time, such "heretical" elements were not necessarily opposed to, but often embodied discursive elements akin to the "mainstream" articulatory practices of politicians. The politician, quoted above, who advocated abandoning half-hearted "participation", also treated D&P as a means of promoting cost-sharing, as attested to by his reference to the importance of "ten percent, twenty percent" of local contributions. This resonated with the "orthodox" discursive practices of political leaders, who tended to uphold "participation" in instrumental terms. The interweaving of these inconsistent elements is also illustrated by the following remark by a political leader affiliated with the CPN-UML, who emphasized the need to allow local people to articulate their aspirations, while at the same time, equating the D&P reform narrowly with people's appealing to the central authority for help.

> "In our view, we have to have a very close link with local communities. If there is no organization at the grassroots level, we do not, or we cannot, know the experiences, the needs of the people. They [the party's grassroots organizations] find all the problems people are facing, and send them to us, the upper bodies. So we always consolidate grassroots organizations."[17]

The discursive field of politicians therefore exhibited a certain degree of "regularity in dispersion" in that the age-old "story-line" of development permeated their articulatory practices irrespective of their subject positions. Even those who did not conform to the "mainstream" discourses tended to depict the flood-affected populace as being at the mercy of the nature, rather than advocating allowing them a voice in influencing the overall direction of the river control policy. This does not mean, at the same time, the "regularity in disper-sion" was far from being immutable. On the contrary, the "regularity" was also vulnerable to other discourses that would reveal the contingency of the basic premises of the traditional "story-line", and was constantly challenged by those counter-discourses. For example, one of such "fragmentary" forces emanated from the emergence of ethnic minorities' rights as one of the political agenda of the national government. This is illustrated by the following remark by one poli-tician who redirected my question about the "participatory" river control policy, to the need of a drastic reform of instituting federalism in lieu of the ongoing unitary state.

> "In my opinion, people's participation is not possible without regional autonomy. How can they participate? If they are given autonomy, they will choose their parliament. And they think 'The state has acknowledged us. We have our parliament. We can choose our leaders.' And then, they will develop their own policies for their area. They will psychologically think that they are now real citizens of Nepal."[18]

As mentioned earlier in this section, the conventional "story-line" of development permeated the discursive practices of political leaders, who thereby tended to relegate "participatory" river control to the technical matter of delivering pre-packaged projects. At the same time, their articulatory practices were vulnerable to competing argumentations that upheld more emancipatory approaches to en-able people to raise wider socio-political issues. As a result, the discursive field of political leaders, who drew upon a stock of varying discourses, constituted an unstable mix of varying and even contradictory elements. A similar picture can be depicted of central bureaucrats concerned with local flood mitigation efforts, whose discursive practices are described in the following section.

## Department of Irrigation (DOI)

Central line agencies were prone to make sense of the D&P agenda, drawing on the "story-line" that portrays villagers as passive recipients of external support for development. This was also the case with the Department of Irrigation (DOI), which was responsible for flood control in the Tarai until 2002. While the De-partment came to direct, in line with the D&P reform, the bulk of its support to "farmer-managed" irrigation systems, it tended to be preoccupied with estab-lishing "water-user associations" to solicit their resources and labor for the con-struction and maintenance of irrigation canals.[19] The DOI was liable to impose detailed pre-packaged support, dictating to farmers the organizational structures of user groups, and their financial plans for repair and maintenance. "People's participation" was therefore equated, within the Department, largely with an instrument to ensure people's acceptance of, and their contributions to the costs of its projects.

Moreover, the DOI bureaucracy generally considered that political leaders tended to form personal cliques to consolidate their power and influence, and to capture and waste development resources for partisan purposes. As a result, the D&P agenda was turned into a form of patronage and favoritism in that it was equated with the general public's appealing to politicians for handouts. The new "participatory" river control policy, according to many of the Department's officials, was the testimony of their propensities to distort development programs with a view to strengthening their personal networks.

It is to be noted, at the same time, that the DOI bureaucracy was not a coherent whole, but encompassed officials with diverse views, whose interactions constantly reproduced and modified the DOI's articulatory practices. There existed some officials who argued for the need to delegate decision-making power to local politicians. It was not plausible to ignore local elected leaders while forming water user groups, given the greater responsibilities accorded to the locally elected leaders under the D&P reform. Some officials even considered it a necessary evil to allow them to compete for "ad hoc" river control resources in response to request from their constituencies because it was part and parcel of the democratization path that the country had chosen to take. The following remark typifies such "heretic" opinions.

> "Democracy has some flaws. We should accept the fact. If you are a politician, you try to please the local people. That is a fact, and the issue is to what extent. 'OK, you try to please the people, but do not violate the country's rules and regulations. Try to live within the limit, but still you can help the people.' Because, as I told you, they live in their districts. We are there only for a short time, and so without involving them, how can we develop their areas?"[20]

Moreover, the instrumental view of participation, referred to above, was also coming under pressure from within the DOI bureaucracy. Toward the end of the 1980s, various government agencies had started adopting policies to promote local self-governing systems for natural resources management, buoyed by the success of community forestry that had contained the rapid deforestation afflicting the country. The resultant prevalence of the "community-based" rhetoric in the country had catalyzed changes in discursive practices within the Department. As a consequence, there existed a growing perception, among the DOI's staff members, of the need to avoid perfunctory involvement of local beneficiaries. This trend is exemplified by the following remark.

> "Do not preconceive any project. OK, if we decide to conduct river control projects, start by making people aware. 'This project is for river control at the local level.' Then we must launch an awareness campaign in the local area. If local people are interested in that project, they are motivated and willing to

participate in the project. And then you can ask people to make a group of their own, to prepare their project, how they want to launch the project, to what extent they want to contribute cash and others [such as in-kind contributions]."[21]

This kind of "fragmentation" within the DOI's discursive field, where the traditional "story-line" of development held sway, emanated from the impossibility of forcing a total closure to the articulatory practices of the Department's officials. Their "mainstream" narratives, described above, in order to provide plausible stories, had to distance themselves from, and were therefore vulnerable to other competing discourses that would reveal the contingency of the "orthodox" accounts. The resultant coexistence of these incoherent elements was also facilitated by their different affiliations and personal backgrounds. Although we should take care not to overemphasize the "social determination of discourse",[22] the articulatory practices of the DOI's staff were influenced, to some extent, by offices or divisions which they were assigned to.[23] Moreover, some of the Departmental staff undertook consultancies for foreign donor agencies, and therefore tended to embrace discourses prevailing in the aid industry.

The intermingling of various competing articulatory practices does not mean that there existed no constancy in the DOI's discursive field. On the contrary, even among those whose articulatory practices embodied "heretical" elements, their concern about local autonomy usually coexisted with their lack thereof. As illustrated by the last two quotations, there existed a certain degree of "regularity in dispersion", in that even those critical of the DOI's continued grip on the development processes still tended to embrace the old "story-line" that depicts local people as in need of guidance from outside agencies. For example, the second narrator, while stressing the need to enhance the capacity of community groups to formulate their own strategies, still limited their role to project implementation, instead of advocating broader public involvement in framing the overall direction of river control.

It is to be noted, at the same time, that the "regularity in dispersion" was not immutable. On the contrary, the "story-line" of development, which served as a common platform for Department's officials with diverse articulatory practices, was ceaselessly contested by other forces that stood at the limits of the "regularity." One source of the destabilization, among others, was the prevailing accusation that the DOI, along with the Roads and Forestry Departments, was the most resistant to giving up their vested interests in controlling the construction and maintenance.[24] This articulation, pointing to the propensity of the Department's staff to seek private financial gains, aroused reactions especially among those who took pride in their commitment to their duties as civil servants. This is exemplified by the following remark, with which one retired senior official argued for the need for the Departmental staff to straighten up their acts, and to serve for the cause of ordinary people.

"It is the people who should construct [river control structures]. Yes. . . . But we should not force people to accept our ideas. Because we have betrayed them several times, if somebody from the Department goes [to a local community with a 'participatory' river control package], they no longer trust us. First, we should sit together with the beneficiaries, and discuss with them what they need. If their needs and our proposals do not match, we should not implement our projects."[25]

The general inclinations of the DOI to treat participation in instrumental terms, based on the conventional "story-line" of development, resonated with those of political leaders. The Department unwittingly formed a coalition with political leaders, to jointly construct the image of flood "victims" as passive recipients of central support. As a result, the "participatory" river control policy was relegated largely to the delivery of technical packages for the flood-affected populace, while emanating an illusion, with its D&P rhetoric, of overhauling the conventional, top-down conduct of development. However, the close examination of the discursive field of the DOI bureaucracy, carried out in this section, reveals that its officials drew on a multiple and contradicting set of discourses, and thus held unstable and incoherent subject positions. This was also the case with political leaders, as described earlier. The policy process of "participatory" river control therefore did not proceed in such a "logical" and purposive manner as stated at the beginning of this paragraph. On the contrary, it entailed constant reflections among political leaders and the DOI's officials, both of whom felt ambivalent about the way the policy was translated into action.

# Decline of the "Participatory" River Control Policy

The performance of district-level river control projects was not encouraging.[26] District-level river control projects usually targeted a limited number of settlements along a river. It was virtually impossible for scattered river control structures to tame a watershed, unlike the situation where there existed continuous dikes along stretches of a river. Small-scale structures at best temporarily solved flooding problems, and often did not withstand the forces for a long period of time. Furthermore, they were only able to shift problems elsewhere, by redirecting river currents to other localities adjacent to the project areas. Despite these drawbacks, the DOI continued to implement district-level river control projects until the newly established Department of Water Induced Disaster Prevention took over the DOI's river control portfolio in 2002. The district-level river control program lingered on although it was clearly tantamount to "words that succeed and policy that fails."[27]

Towards the end of the 1990s, the central government increasingly bypassed the "participatory" river control policy, and instead came to specify where and how to use its budget, instead of sending a blanket authorization to each district. This decline of the "participatory" policy can be attributed to the country's changed political circumstances in the late 1990s. As mentioned above, the "participatory" river control policy was a useful tool for politicians to divert attention away from other possible alternatives to the delivery of small-scale structures, and thus to mask their tendencies to distribute patronage. However, the significance of the policy started to decline as discrepancies arose between the political party in power at the center, and that dominant at the district level. While the central government allegedly rigged the two local elections held during the 1990s, there were frequent changes in the central government, especially during the latter half of the 1990s. As a result, the ruling party at the center did not occupy a majority of the DDC's posts during most of the second half of the 1990s. This caused the policy to decline in its value as a means of distributing patronage, from the viewpoint of national politicians.[28]

This explanation, however satisfactory it may seem, is founded on what Foucault terms a "descending" approach,[29] as described in chapter 2, which looks into who possesses power and what strategies they pursue. This type of analysis overemphasizes the zero-sum nature of power, by attributing it to dominant actors who exercise control over others. Underlying this dichotomous notion of the "powerful" and the "powerless" are the assumptions that individuals' identities are fixed and coherent, and that dominant and less dominant actors are juxtaposed as "baddies" and "goodies." The polarized view is also founded on a "repressive hypothesis" that powerful actors immerse themselves in strengthening and maintaining their influence in society. However, those policy elites concerned with flood mitigation, albeit liable to reduce "participatory" river control to top-down undertakings, did not blatantly seek to inhibit people's active participation. As described above, on the contrary, policymakers were caught in a dilemma over whether local flood mitigation should be relegated to the mere delivery of physical structures.

This study accordingly proposes to undertake an "ascending" analysis,[30] a term coined by Foucault, to examine the micro-power experienced by stakeholders whose perceptions of "participatory" river control were far from being immutable and coherent. The unstable and inconsistent nature of their identities can be attributed to the productive feature of "disciplinary" power. As explained in chapter 2, "disciplinary" power not only inhibits but also facilitates the renegotiations and remolding of the very social norms that it upholds. The prevailing "standard" of equating "participatory" flood mitigation with the distribution of physical structures was ceaselessly contested by other visions that revealed the precariousness of the "orthodox" approach. The discursive fields of policymakers, which resultantly constituted an unstable ensemble of heterogeneous elements, did not confer a fixed, coherent identity on them.

The decline of the "participatory" river control policy therefore did not only emanate from the changed circumstances surrounding policymakers. The "ascending" analysis of the articulatory practices of central policy stakeholders, carried out in this chapter, reveals that the policy decline can also be attributed to the inherently "undecidable" nature of the identities of policy elites. It is to be noted that such an "ascending" approach is not confined to the analysis of "influential" players at the central level, but should encompass the entire spectrum of stakeholders. As explained in chapter 2, "disciplinary" power circulates in society in such a manner as to subjugate both the "powerful" and the "powerless." Accordingly, on the assumption that not only "influential" stakeholders but also their flood-stricken "clients" ceaselessly reflected on and renegotiated the "standard" manner in which local flood control was implemented, in chapters 5 and 6, I undertake an "ascending" analysis of how the micro-power experienced at the village level played its part in the decline of the policy. Before going into the village-level analysis, the following chapter traces the unfolding of the "participatory" river control policy at the district level, and examines how government officials and local politicians in the Bardiya district assimilated the policy into their ongoing local struggles.

## Notes

1. I am simply borrowing the term "D&P" coined by Tendler (1997) as a convenient abbreviation. In this study, "D&P" is not used in the same manner as Tendler who states that policymakers in many countries overestimate the potential of community organisations to prevent local elites from capturing local development resources.

2. David Howarth and Yannis Stavrakakis, "Introducing Discourse Theory and Political Analysis," in *Discourse Theory and Political Analysis: Identities, Hegemonies and Social Change*, eds. David Howarth, Aletta J. Norval and Yannis Stavrakakis (Manchester: Manchester University Press, 2000), 10.

3. Wai F. Lam, *Governing Irrigation Systems in Nepal: Institutions, Infrastructure, and Collective Action* (Oakland: ICS Press, 1998), 9.

4. For the first time in the country's history, the new constitution stipulated that sovereignty is vested in the people of Nepal. The constitution also guarantees standard civil liberties and political rights, such as the rights to speech, assembly, and association, and prohibits discrimination based on caste, ethnicity, and gender.

5. Stacy L Pigg, "Unintended Consequences: The Ideological Impact of Development in Nepal," *South Asia Bulletin* 13, no. 1&2 (1993): 45-58.

6. For example, see Borre, Panday and Tiwari (1994), Dahal (1996), Martinussen (1993), Pandey (1999), and Shrestha (1999).

7. For example, see Meen B. Poudyal Chhetri and Damodar Bhattarai, *Mitigation and Management of Floods in Nepal* (Kathmandu: Format Printing Press, 2001), 49-50. It is to be noted that these alternative measures may not necessarily be feasible in all flood-affected areas. In the case of upper watershed conservation, it is not even certain to what extent floods in the Tarai are exacerbated by human activities to change the surface of the hills and mountains, such as deforestation and farming, as discussed by Blaikie, Cannon, Davis and Wisner (1994, 135-36).

8. Some of the DOI staff members were aware of this flaw, and took the initiative, in the middle of the 1990s, of drafting a "River Control Policy" that would widen the scope of river control efforts. The draft policy paper emphasized the need to diversify the portfolio to include local coping strategies, such as bio-engineering measures. However, the policy paper did not receive due attention within the Department, and was left neglected.

9. Piers Blaikie, Terry Cannon, Ian Davis and Ben Wisner, *At Risk: Natural Hazards, People's Vulnerability, and Disasters* (London: Routledge, 1994), 132-33.

10. Poudyal Chhetri and Bhattarai, *Floods in Nepal*, 50-51. It is to be noted that flood vulnerability is not simply correlated with the extent to which someone's day-to-day livelihood is put at risk. It has to be measured by taking into consideration social networks or contacts one can draw on to cope with emergencies, and to find alternative livelihoods. One should therefore avoid simplistically assuming that the better off are always less susceptible to damage and suffering. At the same time, as argued by Blaikie, Cannon, Davis and Wisner (1994, 10), natural hazards normally leave less severe consequences for the better off, given their larger cash and other reserves which enable them to reconstruct their lives.

11. Hubert L Dreyfus and Paul Rabinow, *Michel Foucault: Beyond Structuralism and Hermeneutics* (Brighton: The Harvester Press, 1982), 196.

12. The CPN-UML came into being in 1991 as a result of the merger of two major groups within the United Left Front, namely a coalition that various community groups formed for the 1990 anti-*Panchayat* movement. While the CPN-UNL occupied the left of the political spectrum, the RPP placed itself at the other end. The NC undertook various progressive measures against land-based feudal exploitation, with socialist ideals, after the electoral victory during the 1950s at the time of the short-lived multi-party system. However, the Congress drifted to conservatism after 1990, by accepting former *Panchayat* leaders into the party.

13. It is to be noted that there existed differences along party lines although they are not referred to in the main text. For example, the CPN-UML regarded "working class people" and "peasants" as the vanguard of socio-economic progress, in contrast to the NC and the RPP, both of which tended to lump "people" together and patronized them as recipients of development. At the same time, even the CPN-UML drew on the conventional "story-line" to patronize local governing bodies and people. This is well illustrated by the accusation that the CPN-UML (1998, 17) directed at the NC of carrying out decentralization by halves but argued for the need to "deliver them the fruits of the development", rather than the importance of unleashing grassroots political activism for social change. The CPN-UML's vision, albeit its alleged commitment to abolishing "semi-feudal" society, was similar to that of the NC, in their top-down perspectives.

14. Since the first general election was held in 1991, Nepal had as many as nine governments by the end of the decade. The NC formed a majority government from 1991 to 1994, and from 1999 to 2002, while the CPN-UML formed a minority government from 1994 to 1995. The RPP was a coalition partner in the NC-led government from 1995 to 1997, and in the CPN-UML-led government in 1997, and again in the NC-led government from 1997 to 1998.

15. The "non-fixity" of politicians' discursive practices also emanated from those associated with academia and NGO circles, who assumed ministerial and other political appointees' posts (such as the memberships at the National Planning Commission). There was a tendency, among academics and NGO personnel, to criticize the D&P initiative for

failing to grant autonomy to local bodies. This propensity served to destabilize the discursive field of politicians, a majority of whom were less critical of the D&P program.

16. Excerpt from my interview with a politician affiliated with the CPN-UML in August 2000.

17. Excerpts from an interview of July 2000.

18. Excerpts from my interview with a political leader from the NC in July 2002.

19. Lam, *Irrigation Systems*, 42.

20. Excerpts from my interview with one Divisional Engineer of the DOI in July 2002.

21. Excerpt from my interview with an official of the DOI's Management Division in January 2001.

22. Norman Fairclough, *Discourse and Social Change* (Cambridge: Polity Press, 1992), 65.

23. Officials at the Management Division, for instance, were generally more attentive to the operation and maintenance of the DOI support, and were more inclined to propose to do away with perfunctory involvement of beneficiaries.

24. ADDCN, *Decentralization in Nepal: Prospects and Challenges (Findings and Recommendations of Joint HMG-Donor Review)* (Kathmandu: ADDCN, 2001), 22; Lam, *Irrigation Systems*, 196.

25. Excerpts from my interview with a former Deputy Director-General of the DOI in August 2000.

26. Poudyal Chhetri and Bhattarai, *Floods in Nepal*, 52.

27. Murray Edelman, *Political Language: Words That Succeed and Policies That Fail* (New York: Academic Press, 1977).

28. For example, when the NC formed a coalition government in 1998 and a subsequent one-party government in 1999, about eighty percent of the DDC chair's posts had been occupied by the CPN-UML. As a result, the NC government resorted to sidelining the DDC chair in the management of line agencies' portfolios, including the local river control program.

29. Michel Foucault, "Two Lectures," in *Power/Knowledge: Selected Interviews and Other Writings 1972-1977 Michel Foucault*, ed. Colin Gordon (New York: Harvester Wheatsheaf, 1980), 99-100.

30. Foucault, "Two Lectures," 99-100.

# Chapter 4
# Localized Meanings of "Participatory" River Control in Bardiya

This chapter analyzes how policymakers in the Bardiya district interpreted and contested the meaning of the "participatory" river control policy. For this purpose, it focuses on the two main groups that played key roles in shaping the policy process at the district level, namely political leaders at the district level, and officials at the District Irrigation Office (DIO). Before analyzing the policy process, the first section of this chapter provides an overview of the district's history, to highlight how the corvée tradition called *begaari* had originated in the colonization of the area by emigrant highlanders, *Pahadis*, in the nineteenth century. The *begaari* obligations for such public works as road construction, irrigation canal maintenance, and river control, had ever since fallen largely upon the original settlers, *Tharus*, most of whom continued to depend on *Pahadi* landlords for their livelihoods. As a result, the corvée practice was long central to power struggles in Bardiya, which manifested themselves most clearly along the line of ethnicity.[1]

In Bardiya, the "participatory" river control policy was translated into action with the mobilization of unpaid labor in line with the *begaari* tradition. As stated in the second section of this chapter, the conventional "story-line" of development exerted a strong influence in the discursive practices of district-level politicians and officials at the DIO. In line with the "story-line" that portrays villagers as passive recipients of external help, both actors found a common ground in treating the flood-affected populace as victims of nature, and in equating flood control with the construction of physical structures. "Participation" was relegated to being an instrument to get people, through their labor contributions, to execute pre-determined projects. At the same time, this does not mean that local leaders and officials at the DIO did not hesitate to proceed with the corvée-based construction of flood control structures. On the contrary, their outlook on "participatory" river control wavered owing to the penetration into their discursive fields, of various counter-discourses, such as those pointing out the importance of redressing the injustice embedded in *begaari*, and the need to broaden the scope of flood mitigation efforts. The policy-process of "participatory" river control in Bardiya therefore did not unfold in such a "logical" and linear manner as described above. On the contrary, it entailed constant reflections and renegotiations among local actors.

As described towards the end of chapter 3, in the late 1990s, the "participatory" river control policy came to be increasingly ignored by the central government. This was also the case with the Bardiya district, where there were hardly any *begaari*-based river control projects after 1996. This chapter concludes with discussions of how the case of Bardiya can possibly be explicated with the "descending" and "ascending" perspectives, touched upon in chapter 3. Drawing on a conventional "descending" analysis, it can be surmised that the ruling party in the center intended to bypass elected politicians from the opposition camp in Bardiya, in order to prevent them from using the river control portfolio for partisan purposes. At the same time, given the unstable and inconsistent nature of the identities of local politicians and government bureaucrats, one can also make sense of the decline of the policy from an "ascending" perspective that focuses on the dilemmas in which they were caught over whether or not they should carry on with *begaari*-based river control projects.

# Background of the Policy Process in Bardiya

Nepal is divided into three physiographic zones, the mountains, the hills, and the Tarai plain. The Tarai, which is an extension of India's Gangetic plains, stretches along the southern frontier adjoining India. The Bardiya district is the third Tarai district from the western border with India. In the district, the original settlers called *Tharus* historically provided unpaid labor for public works. Conflicting meanings attached to the corvée tradition, which was long central to power struggles among the local populace, also exerted an influence on the district-level policy process of "participatory" river control. This section first provides an overview of the history of Bardiya to elucidate the origins of the corvée practice. It then explains the historical corvée tradition and the divergent interpretations accorded to it by residents in the district.

Foucault points out that "the history of government . . . is interwoven with the history of 'dissenting counter-conducts.'"[2] The accounts given in this section, however, leave out "dissenting counter-conducts", with little regard paid to the role played by the public in the unfolding of historical events. Underlying the seemingly unquestionable and lucid stories given in this section is a "repressive hypothesis", explained in chapter 2, which attributes power only to political elites, but not to their "clients." As explained in chapters 5 and 6, local politics in Bardiya was played out through a web of complex micro-power struggles, in which even the general populace exercised leverage over its proceedings. This section purposefully focuses on the unjust exercise of power by dominant actors, with a view to elucidating the background of the myriad of responses and reactions by the general public. As implied in the title of the following section, this section therefore offers "a political history", not "the history" of the Bardiya district.

## "A History" of the Bardiya District

Bardiya and three other neighboring districts were annexed to Nepal in 1861, in return for the military support that the Nepalese state extended to the British Empire in subduing an insurgency in northern India. The then ruler of Nepal subsequently granted tracts of land as rewards to their loyal courtiers and generals who were high-caste *Pahadis* of hill origin. During the second half of the nineteenth century and the early twentieth century, Bardiya was literally "*birta*", namely the land granted to the country's ruling families by the state. As a consequence, the percentage of landholding by *Pahadis* landlords increased from thirty-two percent in 1910 to sixty-three in 1964.[3] No other district in Nepal had as large a *birta* as Bardiya, and the predicament of peasants in Bardiya was particularly severe.[4]

Baridiya and surrounding areas had historically been marginal to the economy of the northern India states, and had mostly been a sparsely inhabited jungle. Newly settled *Pahadi* landlords were required to bring their land under cultivation, and to collect land taxes. In return, they were given extensive administrative and judicial power to govern their estates, and were therefore able to arbitrarily set rents, and exact unpaid labor and services from the peasants. Bardiya was historically thinly populated mostly by *Tharus* and a limited number of Hindu caste groups. The colonization by the state impacted on preexisting settlers in two ways. First, a majority of original inhabitants became tenants or laborers for new landlords, and were rendered economically dependent on them. Second, new settlers reduced *Tharus* into a lower status of enslavable *Matwali* by introducing the caste system,[5] and exacted their unpaid labor and services. As a result, local peasants had to suffer from the double burden of high rents and corvée obligations called *begaari*, the latter of which is discussed in the following section.

The historical take-over of land by newcomer settlers rendered many *Tharus* destitute, who often fell into debt and became bonded laborers called *kamaiyas* for their new landlords. In such cases, all family members were required to provide labor and services for their landlords, in exchange for a minimal amount of grain, barely enough for their survival. A *kamaiya* family was obliged to carry out both agricultural and domestic tasks as demanded by their landlord.[6] On the other hand, *Tharu* peasants who were endowed with the necessary labor, equipment, and oxen enter into tenancy agreements, under which they took on full responsibility for cultivating the land of their landlords, and were usually entitled to retain half the yield. Unlike *kamaiyas*, tenants were not on call twenty-four hours a day, and were in principle not required to provide unpaid labor and services, either. However, many of their landlords took advantage of their lack of alternative livelihoods by manipulating the terms of their tenancies.[7]

Not all *Tharus* were landless or small farmers, but a fraction of them were well-to-do owner-cultivators. There existed, however, a common feeling of predicament among *Tharus*, either well-off or poor, towards the hill-dominated administration and culture. Hindu high-caste *Pahadis* from the hills constituted a powerful force, not only in economic, but also in political terms. *Pahadis* drew on their education, and their caste and kinship affiliations with government leaders and functionaries, to assume most of the elected posts of the local governments, and also of community groups for government-supported programs. Moreover, the influx of *Pahadis*[8] had caused scarcities of cultivable land, pastures, and forests, thus exerting pressure on the limited local resources. This also meant that *Tharus* were also deprived of their resources to conduct their rituals, thereby bringing about a crisis in their cultural identity.

In Bardiya, just as in other Tarai areas, resettlement and land reform programs tended to favor *Pahadis*, and failed to alter the skewed distribution of land that had originated at the beginning of the *Pahadi* rule in the nineteenth century. High-caste *Pahadis* continued to dominate the politics of Bardiya in the sense that the elected posts of the District Development Committee (DDC) or district-level committees of political parties were mostly occupied by them, so were the leadership positions in government-supported programs. As noted at the beginning of this section, the accounts given in this section shed light only on the unjust exercise of power by the Nepalese state and migrant *Pahadis*. In Bardiya, however, politics was played out through a web of intricate micro-power struggles, in which dominant players did not necessarily seek to suppress *Tharus* at all times, and even the oppressed exercised some leverage over the evolution of political histories. Before elucidating such "dissenting counter-conducts" in the latter half of this chapter, and also in chapters 5 and 6, the following section deals with a prominent feature of the *Pahadi* domination, that is the age-old tradition of corvée practices.

## *Begaari* Tradition for Public Works

In many parts of the Bardiya district, public works were long undertaken with community labor contributions. This tradition, originating in the advent of the *Pahadi* domination in the nineteenth century, was central to power struggles among the local populace because corvée practices reflected ongoing asymmetrical social relations. In Bardiya, *Tharus* had been engaging in rain-fed agriculture before their land was taken over by *Pahadi* landlords who were given responsibility for the collection of land taxes. In order to boost agricultural production, new landlords set about the development of irrigation systems. *Tharus*, who had been relegated to the status of the hired hands of *Pahadi* landlords, were entrusted not only with the construction of waterways, but also with the maintenance and repair of irrigation canals. Moreover, new landlords were given, as stated in the preceding section, extensive administrative and judicial power to govern their estates, and also exacted unpaid labor from *Tharus* for other public works, such as the construction and maintenance of village roads and bridges.[9]

Prior to the colonization by *Pahadis*, there had been a communal labor system, under which *Tharus* contributed their voluntary labor for collective actions, including rituals and agricultural activities.[10] The term *"begaari"* derived from this practice. *Tharus* had long engaged in *begaari* through their indigenous organization called *khel*, that was represented by male heads of all the households in each unit of mutual assistance. Key posts traditionally included a *badghar* who led each *khel*, a *chowkidar* who was the messenger, and those who played key roles on religious occasions. The *khel* served as a convenient tool for *Pahadis* in exacting unpaid labor from *Tharus* for new public works. *Pahadi* rulers entrusted the *badghar* to take the lead in organizing labor contributions, and the *chowkidar* to give out messages and instructions to member households. As a result of this appropriation by *Pahadis*, the term *"begaari"* came to have negative connotations among the *Tharu* population.

As stated in the preceding section, in Bardiya, despite the land reform during the 1960s, most agricultural land continued to be owned by *Pahadis*. Moreover, *Pahadis* remained as a powerful force both economically and politically, and continued to rely on the unpaid labor of *Tharus* for public works. As a result, the age-old tradition of *begaari* lingered on in Bardiya in the 1990s, and its major features remained largely unscathed in many places in the district, namely those who actually worked in the field were responsible for maintaining irrigation canals and other infrastructure. *Pahadis*, most of whom used *Tharu* tenants or *kamaiyas* to cultivate their land, were thereby exempted from the *begaari* requirements.

*Begaari* was applied not only to local self-help activities, such as the maintenance of intra-village roads, but also to externally supported projects, including river control projects. With the renewed emphasis on decentralization and participation (D&P), the government was channeling a larger portion of its budget to villages during the 1990s. As a result of the growing availability of governmental funding for village public works, *Tharus* came to be compelled to undertake a wider range of *begaari*. In addition to the maintenance of basic infrastructure, such as irrigation canals, and intra-village roads and bridges, *Tharus* were increasingly required, in the 1990s, to engage in new types of public works, such as the construction of school buildings and flood control structures, which required such external inputs as cement and steel. Even basic infrastructure, such as canals and village trails, came to be upgraded with government funding using non-local materials.[11]

At the same time, the country's 1990 political change, from the monarchical *Panchayat* rule to the multi-party system, opened up space for *Tharus* to express their grievances in public. In the 1990s, *Tharus* were increasingly resorting to overt protest against the inequalities entailed in the *begaari* tradition. *Tharu* peasants generally resented the *begaari* practice as an instrument that *Pahadis* had long used to dominate *Tharus*. *Pahadis* historically preempted governmental programs, using their connections with functionaries, and had also manipulated the land reform programs in which *Tharus* had been deceived into parting with their farmland. As a consequence, *Tharus* continued to be forced to work as laborers or tenants for *Pahadi* landlords, and to provide unpaid labor for village public works even in the 1990s. There existed widespread suspicion that *Pahadis* embezzled the government's resources in collusion with officials, while covering the misused funds with free labor contributions. *Tharu* peasants also described *begaari* as unjust on the grounds that it compelled them to forgo their own work for days, in spite of the fact that they were living hand-to-mouth.

In general, on the other hand, *Pahadis* justified the *begaari* tradition by projecting themselves as benevolent benefactors helping uneducated *Tharus* to improve their living standards. Since *Pahadis* first arrived from the hills in the nineteenth century, to propagate irrigation in place of rain-fed agriculture, *Pahadi* migrants always led *Tharus* in the development of the village, of which the corvée practice was always an integral part. In this way, the traditional "story-line" of development prevailing in the country, which constructed the polarized scale of progress and backwardness, was mobilized by *Pahadis* to contrast themselves and *Tharus*. The D&P rhetoric, holding sway in Nepal to uphold participation as an instrument to solicit people's "voluntary" contributions, was also relied on by *Pahadis* in standing behind the traditional corvée system, as a tool to enhance the cost-effectiveness of village public works.

As noted above, these ostensibly unquestionable and reasoned descriptions did not constitute the essentials of the articulatory practices of *Pahadis* and *Tharus*. On the contrary, their perceptions of *begaari* were fluid and even fragmented. As explained in chapters 5 and 6 dealing with the village case studies, villagers, both *Pahadis* and *Tharus*, incessantly readjusted their subject positions, owing to the precariousness of the above articulatory practices concerning the corvée practice. Neither of the assertion for or against the *begaari* tradition were able to stay immune from the counter-discourses, nor was either able to accord fully constituted identities to villagers. Analysts should therefore avoid surmising a likely course of social interactions abstractly from the above rather reified representations, and instead make careful empirical inquiries into everyday dilemmas facing villagers. This point is dwelt upon in chapters 5 and 6.

# "Regularity in Dispersion" of District-level Actors' Stances

In the Bardiya district, the "participatory" river control policy was translated into action by drawing on the local corvée tradition. Just as the policymakers at the central level, political leaders in Bardiya and officials at the District Irrigation Office (DIO) generally coalesced around the "story-line" of development prevailing in Nepal, namely that some people are inherently more advanced than others. This provided grounds for them to proceed with the use of the local corvée practice that fell inequitably on *Tharus*. At the same time, neither local politicians nor staff members at the DIO blatantly sought to reduce the policy to the corvée-based construction of river control structures. On the contrary, since they felt ambivalent about the imposition of pre-packaged projects, the district-level policy process entailed constant reflections among local leaders and officials. This section looks into how the seemingly fixed definition of the "participatory" policy was ceaselessly renegotiated in the Bardiya district, and thus led to the continual struggles over the policy's localized meanings.

## Political Leaders in Bardiya

Politicians in Bardiya generally stood behind the *begaari* tradition, partly owing to the *Pahadi* dominance in the local political circles. Most local leaders who assumed posts at the DDC, or on district committees of major political parties, namely the Nepali Congress (NC), and the Community Party of Nepal-Unified Marxist-Leninist (CPN-UML), and the *Rashtriya Prajantra* Party (RPP),[12] were *Pahadis*. *Pahadi* politicians were prone to support the corvée practice since their dominance originated from the asymmetrical *Pahadi-Tharu* relations founded on *begaari*. At the same time, we should not take for granted the "social determina-tion of discourse",[13] namely the presumption that the ethnic affiliation of re-spective political leader determined how each made sense of the corvée practice. Even *Tharu* politicians who presumably deplored their kin's or neighbors' plight, usually assimilated themselves to the corvée practice. *Tharu* leaders, as repre-sentatives of the mainstream parties dominated by *Pahadis*, rarely confronted the "orthodox" articulatory practices of district politicians.

The polarized scale of advancement and backwardness, that is the central premise of the conventional development "story-line" in Nepal, was embraced by local politicians. They normally projected *Pahadi* migrants as benevolent bene-factors, who introduced irrigation systems and other development into the backwaters inhabited by *Tharus*. The use of the *begaari* tradition, which *Pahadi* migrants had initiated in order to guide uneducated *Tharus* to development, was therefore an unquestionable and matter-of-course option for river control. The D&P rhetoric was also appropriated by political leaders who used the term, *jana sahabhagita* or people's participation as a euphemism for corvée practice. In this way, local politicians drew on the traditional "story-line" of treating development as exogenous to villages, and promoted river control as an instrument to per-petuate their dominant position in which *Pahadis* exacted unpaid labor from *Tharus*.

Local politicians' discursive practices, however, were not necessarily com-patible with the conventional "story-line" of development. They constituted an unstable ensemble of incoherent and contingent elements because, like any other discourses, the "story-line" was vulnerable to counter-discourses that were equally plausible and would potentially reveal its precariousness. Local leaders' articulatory practices were therefore neither homogeneous nor immutable, but were fluid, multiple and contradictory.[14] While there existed a certain degree of "regularity" in their discursive practices, it is crucial to take note of the "dispersion." The conventional "story-line" served as the backbone of the "regularity in dispersion" in the discursive field of local politicians, but did not constitute an essential feature of their articulatory practices.

Such counter-discourses which reveal the contingency of the "regularity" included an argumentation that small-scale structures built by ordinary people could not withstand the forces of floodwaters all the time, but would at best be a temporary solution to flooding problems. The rationale of *begaari* also dwindled away because it would force *Tharus* to forgo their own work for days in spite of their hand-to-mouth way of life. These types of counter-arguments were equally tenable, and thus made many political leaders feel ambivalent about whether or not they should routinize "people's participation" in public works. This point is illustrated by a politician who wavered between his discontent with the way "modern" technologies had been eroding the tradition of "people's participation", and his admission of the limitation of the *begaari* practice.

> "Government engineers and overseers use new technologies, and they are government employees. As a result, local people started questioning why they should work without pay. They are busily occupied with farming, and have little time [to spare]. It is also true, when we use contractors to build [flood control] structures, the quality is good because they have knowledge. Small-scale structures can be built by local people, and large-scale projects should be made by [government] engineers and contractors."[15]

Another source of "fragmentation" in local leaders' discursive practices emanated from their assertion that what was at stake in flood control was the "public" interest, that is it would benefit everybody, regardless of their affiliations, ethnicity, or classes. By stressing the "public" nature of flood mitigation endeavors, political leaders attempted to create an impression that "everybody" would join hands in constructing river control structures. This seemingly praiseworthy story was vulnerable to counter-arguments that problematize the inequality historically entailed in the corvée tradition. The precariousness of the assertion manifested itself as the growing tendency in the late 1990s in Bardiya, to request various actors to provide a helping hand, such as political parties, government agencies, local governing bodies, and non-governmental and civil society organizations. The project taken up in chapter 6 is a case in point, in which river control structures were constructed to contain riverbank erosion threatening a nearby market town, with the unpaid labor contributions of a range of actors. This dilemma facing local leaders, having to mobilize local elites and even public servants, in order to carry through the *begaari* tradition, is illustrated by the following remark by a member of the project's management committee.

"Why I was involved was that I am also a social worker, although I am not an elected political leader. . . . Not only government officials, but also social workers took part in order to encourage the general public. . . . Everybody provided help. Even the SP [Superintendent] of the District Police himself filled sandbags. While constructing the river control structures, we did not differentiate anybody, but involved everybody including the CDO [Chief District Officer], the Mayor, the DDC chair."[16]

This trend among local leaders, towards loosening their grip on the corvée practice, was prompted by the ever-intensifying resistance by *Tharus* to the *begaari* practice, which compelled political leaders to show more and more leniency towards their plight.   As explained in chapters 5 and 6 on the two village case studies, the 1990 political change had opened up space for *Tharus*, who increasingly challenged both in public and through covert forms of resistance, the inequalities entailed in *begaari*. Moreover, there were a growing number of *Tharus*, albeit a minority in number, who assumed posts at the DDC, or on district committees of the major political parties. Many of these *Tharu* leaders were either suffering from the unequal practice themselves, or were seeing their kin and neighbors experiencing hardships. As *Tharus* gradually expanded their political influence, political leaders were increasingly compelled to abandon high-handed approaches to the implementation of *begaari*. The following statement in which one leader lamented the growing difficulty, and at the same time admitted the need to engage in dialogue with *Tharus*, captured the dilemma facing local political leaders.

"Before [under the monarchical *Panchayat* rule that had lasted for several decades until the 1990 political change], people were afraid of their landlords. So a lot of work was done [with labor contributions by *Tharus*]. Nowadays [with the advent of the multi-party democracy], they started standing against us, and we must make them understand [that they have to do so]."[17]

As described at the outset of this section, political leaders in Bardiya generally endorsed the application of the local corvée tradition to the "participatory" river control policy. The reasoning behind equating people's participation with *begaari*, however, could not hold but was challenged by counter-discourses that point to the contradictions inherent in the "orthodox" accounts. Local leaders were thereby caught in a dilemma over whether to continue drawing on *begaari*, or to avoid relying on the unpaid labor of Tharus. Such feelings of ambivalence were also prevalent among officials of the District Irrigation Office (DIO), who generally did not favor allowing villagers a voice in the planning and implementation of river control projects, drawing upon the conventional "story-line" that treats villagers as recipients of development brought in by external benefactors. At the same time, the DIO increasingly accepted the necessity of opening up its hierarchies to local actors, as described in the following section.

## District Irrigation Office (DIO)

Officials at the DIO in Bardiya,[18] just as those at the DOI in the capital, generally expressed discontent with the "participatory" river control policy, which compelled it to run its flood control portfolio under the leadership of the DDC chair. According to staff members at the DIO, however, local political leaders were prone to pursue a short-term, parochial goal of maintaining and strengthening their support bases, at the expense of the overall welfare of the flood-affected populace. The DDC chair was not able to look beyond his favored areas, and tended to come up with structures that would protect the project sites at the expense of other areas. From a technical point of view, this led to serious design flaws which could even be detrimental to the long-term safety of the protected areas, thus preventing the government from deriving maximum benefit from its investment.

At the same time, the discursive field of the DIO was neither static nor homogenous, but constituted an unstable ensemble of incoherent elements, drawn from a stock of competing discourses. The above accusation concerning politicians' tendency to focus on distributing patronage was vulnerable to the D&P rhetoric, according to which the restoration of the multi-party system in 1990 had created a "political market" to hold local leaders accountable to the general public. Local political leaders, to whom the Nepalese people were able to appeal for better welfare and livelihood, were playing an important role in nurturing positive ties between development and democracy. In reality, moreover, the rise of the D&P agenda of the country increasingly required the DIO to involve local politicians in its activities. The following remark by a former Chief of the DIO in Bardiya attests to a growing albeit somewhat unwilling acceptance within his office that local leaders had a role to play.

> "They [local politicians] give projects to their own areas, for their own people. They are only interested in saving their own villages, not in safeguarding the river [in totality]. But they also have an advantage. They know the reality. Government officials do not go [around areas of their jurisdictions], but they [local leaders] visit everywhere in Bardiya. They have better ideas of local areas. The downside is that they favor their own areas."[19]

The "P part" of the D&P discourse was generally not accepted by staff at the DIO, either, who usually embraced the conventional "story-line" of development, and regarded the flood-affected populace as passive victims of nature. The DIO resisted allowing beneficiaries to participate in river control activities because villagers, so the story went, were struggling to make ends meet, and had no time to think over what actions they should take. People did not seriously consider the consequences of their reckless behaviors of encroaching on riverbanks to engage in cultivation or to pasture their livestock. According to officials at the DIO, in no way were people able to see flooding in perspective, nor to visualize optimal solutions to forego their narrow interests. Flood control should be led by district engineers, who were technicians capable of prescribing for flooding problems.

At the same time, the DIO's reluctance to work with beneficiary groups was also contested by other interpretations. The DIO, in line with the country's "participatory" irrigation policy, had come to direct the bulk of its support to farmer-managed irrigation systems. This resulted in an increasing realization within the DIO that participation would also be imperative for the successful implementation of their river control projects. The above "orthodox" accounts, moreover, were prone to be infiltrated by various counter-discourses. A natural event would not turn into a "disaster", unless people encroached on flood plains. Flood mitigation should therefore include efforts to discourage people's reckless behavior and to prepare them for potential damage and suffering. An assertion about the need to solicit the participation of the flood-affected populace and to raise their awareness was bound to seep into the "story-line" that relegated local people to the role of passive victims.

In this way, the articulatory practices of the DIO became increasingly fluid, and came to reflect the ambivalent feelings of its staff members in relation to the D&P rhetoric that gradually penetrated their portfolio. At the same time, the DIO was still liable to revert to the conventional development "story-line" prevailing in the country that regards rural areas as backwaters. For example, in accordance with the instruction from the center, in 2000, the DIO took an initiative to organize flood-affected communities, to regenerate vegetative covers on the riverbanks. Even under this "participatory" program, community groups were assumed to be passive entities to accept the preconceived plan to plant trees and shrubs. Officials at the DIO, just as their central counterparts at the DOI, treated "participation" as an instrument to enlist beneficiaries' acceptance of, and to solicit their contributions to preconceived activities. Therefore, although the D&P rhetoric was gaining ground in the DIO's discursive field, its staff tended to treat it within the framework of the conventional "story-line."

However, this "regularity in dispersion" was far from being immutable. On the contrary, the "regular" element was ceaselessly contested by other competing discourses that had been excluded from its discursive formation. The DIO's articulatory field increasingly embodied "heretical" elements, namely advocacies of more "emancipatory" approaches that would do away with the "orthodox" practice of getting the local populace to follow pre-determined activities. These elements diverged from the "regularity" in that they stressed the need to focus on enabling local leaders and the general public to upgrade their capacity to manage their own development. The following statement by one engineer of the DIO illustrates how the "regularity" in the Department's articulatory practices was be-ing challenged by other more "liberating" discourses.

> "There is a feeling of ownership among the people after the construction [of the structure, when we adopt a "participatory" approach]. For instance, there are many cases where a structure is damaged, but again people go and try to repair it on their own. . . . There is a way out [even with regard to the partisan tendency of local leaders]. If you contact only one person, then, you will give support to that person, or maybe people of his own party. But if you involve people from different political parties, he cannot misuse his position."[20]

In this way, the discursive practices of the DIO's staff members were increasingly fragmented by their admission of, and their antipathy to the desirability of in-volving local politicians and local communities. Such "fragmentation" within the DIO's discursive field, where the traditional "story-line" of development held sway, resonated with the incoherent nature of the articulatory practices of local politicians, described in the preceding section. Though the two actors in Bardiya found a common ground in equating "participatory" river control with the corvée-based construction of physical structures, therefore, the district-level policy process entailed ceaseless reflections on the way the policy was translated into action. The "story-line" of development prompted both stakeholders to relegate river control to the technical matter of protecting "victims" through pre-packaged engineering solutions. At the same time, it was not possible to force a total closure to the discursive fields of the two actors because their "orthodox" articulatory practices were vulnerable to counter-discourses that would reveal their contin-gency. The following section considers how one should make sense of the decline of the policy towards the end of the 1990s, in the light of such contested nature of the district-level policy process.

# Decline of "Participatory" River Control in Bardiya

As mentioned in chapter 3, towards the end of the 1990s, the central government increasingly specified the locations and types of structures and the use of con-tractors, in lieu of a blanket authorization for each district. This trend also manifested itself in the Bardiya district. In the latter half of the 1990s, there were no projects in Bardiya that followed the "participatory" river control policy, ex-cept for some emergency works that required local labor contributions, given the absence of prior budget allocations.[21] One such exception was a project, to be discussed in chapter 6, which was undertaken in 1999 in response to the rapid riverbank erosion that posed an imminent threat to a market center. On the other hand, with regard to the regular budget allocation for the Bardiya district, the central government imposed a pre-packaged river control project.[22]

As also stated in the final section of chapter 3, this move towards centralization can be attributed to the discrepancies among political parties assuming power in the center and dominant at the district level. The two local elections held in the 1990s were allegedly rigged by the central government. However, given the frequent changes in the central government in the latter half of the 1990s, the ruling party did not enjoy a majority of the elected posts of the DDC and the Village Development Committee (VDC). This was also the case with Bardiya where the CPN-UML occupied, just as in most other districts, the bulk of the seats in the local bodies in the late 1990s. The NC was in power at the center from 1997 to 2002, initially as a coalition partner but subsequently in its own one-party governments. It can be argued that the Congress party imposed centrally directed projects on Bardiya to avoid allowing elected politicians from the CPN-UML to use the river control portfolio for partisan purposes. The NC intended to deliver the resources directly to supporters of the party at the grassroots.

This description, however reasoned and unquestionable it may seem, does not elucidate the entirety of the background resulting in the decline of the policy. Underlying this "descending" analysis, a term coined by Foucault,[23] is an assumption that local political leaders acted as middle persons who sought to control and distribute goods and services handed down by the center. The above explanation focuses merely on the zero-sum nature of power, and attributes power only to political leaders at the center as well as in Bardiya, who presumably strove to exercise control over ordinary people in the district. Such a dichotomous notion leads us to surmise that individuals' identities are fixed and coherent, and that "powerful" and "powerless" are juxtaposed as "baddies" and "goodies", drawing on a "repressive hypothesis."

As described above, however, the identities of politicians in Bardiya were far from being stable and monolithic. On the contrary, local leaders wavered between their propensities and their disinclinations to equate "participatory" river control entirely with *begaari*-based endeavors of constructing physical structures. This was because other competing discourses, such as those pointing to the inequalities entailed in the corvée tradition, and to the need to undertake more holistic efforts beyond the mere construction of small-scale structures, seeped into the foundation of the above "orthodox" accounts. The downward trend of the "participatory" river control policy can therefore be attributed to the inherently ambivalent nature of the identities of local leaders, just as it emanated from the fluid identities of national politicians who, as described in chapter 3, did not blatantly seek to treat the policy as an instrument for patronage distribution.

This study, in making sense of the decline of the "participatory" river control policy, advocates undertaking what Foucault terms an "ascending" analysis[24] that examines the micro-level, inconspicuous power dynamics that rendered the subjectivities of stakeholders multiple and inconsistent. As explained in chapter 2, "disciplinary" power, permeating society in such a manner as to operate on both "powerful" and "powerless" actors, not only inhibits but also facilitates the remolding of the very social norms that it upholds. This self-reflexive nature of "disciplinary" power is attributable to the susceptibility of a discourse, which supposedly confers a fixed and coherent identity on a social actor, to other competing forces that stand at the limit of the discursive order. Accordingly, the norm prevailing in Bardiya of equating "participatory" river control with the corvée-based construction of physical structures was contested by other competing interpretations, as illustrated in this chapter, describing the contested and resultantly unstable and incoherent nature of the discursive practices of local politicians, as well as of staff member at the DIO.

An "ascending" analysis should not be confined to "influential" policymakers, but should also include flood-stricken villagers who were also subjugated to, and resist and respond to the "disciplinary" power equating local flood control with *begaari*-based endeavors. The following two chapters, which study two specific villages, focus on the micro-power experienced at the village level, given the hypothesis of this study that the entire spectrum of stakeholders, including policy "clients", exerts some leverage over the policy process. Chapter 5 deals with one village where river control projects were undertaken in 1993, 1994 and 1996, while chapter 6 takes up river control projects that were implemented in another village in 1999, 2000, and 2001.

## Notes

1. Ethnic consciousness arises out of the ensemble of identities formed around a common history, language, territory, and culture. Ethnic identity is therefore a shifting category and is difficult to capture. In the case of this study, a sense of shared history plays an important role, in line with the result of a study by Guneratne (1998) on the construction of ethnic categories in Nepal. The *Tharu* identity and the *Pahadi* identity, which were a main source of village power contestation, were nurtured from the history of their relationships to the state and their resultant positions in Nepalese society.

2. Foucault 1978 cited in Colin Gordon, "Governmental Rationality: An Introduction," in *The Foucault Effect: Studies in Governmentality*, eds. Graham Burchell, Colin Gordon and Peter Miller (Chicago: The University of Chicago Press, 1991), 5.

3. *Bhumishdhar Bibhag, Bardiya Jillama Bhmishdhar* (Kathmandu: *Sree Panchko Sarkarko Chhapakhana*, 1966), 5.

4. Suresh Dakhal, Janak Rai, Dambar Chemchong, Dhruba Maharajan, Pranita Pradhan, Jagat Maharajan and Shreeram Chaudhary, *Issues and Experiences: Kamaiya System, Kanara Andolan and Tharus in Bardiya* (Kathmandu: SPACE, 2000), 39.

5. According to the *Muluki Ain* or the Civil Code enacted by the Nepalese state in 1854, the population was categorized into four broad caste groups, namely *Tagadhari* (Wearers of the Holy Cord), *Matwali* (ethnic groups), the impure but touchable group, and the impure and untouchable group. *Matwali* was further divided into two categories, namely non-enslavable and enslavable.

6. There were also families that were not indebted to their landlords, but worked as *kamaiyas*, given no other livelihood opportunities. Alternatively, only one person from a

family entered into a *kamaiya* contract with a landlord, under which the laborer similarly was required to perform various tasks for a meager wage.

7. Tenant farmers were often obliged to undertake, just as *kamaiyas*, such tasks as carrying crops to mills, collecting firewood and fodder from the jungle, preparing for special rituals, and plastering and cleaning the houses. Some landlords even forced their tenants to cultivate a small plot of land without pay, or to provide their children as live-in servants.

8. In Bardiya, the in-migration of *Pahadis* accelerated especially in the 1960s. A major event that facilitated the influx of migrants from the hills was a rush of land title transfers that *Pahadi* landlords resorted to, to bypass the 1964 land reform program. Just before the reform was implemented, many of them sold portions of their land to other high-caste *Pahadis* who were seeking better livelihood opportunities in the Tarai. The ratio of the *Tharu* populace in Bardiya, which was registered at eighty percent in 1950, decreased to fifty-five percent in 1991, according to the government's population census.

9. *Begaari* also encompassed tasks that *Tharu* laborers or tenants performed on the private properties of their landlords on an unpaid basis, such as those described in the preceding section. This section, however, focuses on *begaari* for public works projects, a subject of central concern in this study.

10. Dakhal, Chemchong, Maharajan, Pradhan, Maharajan and Chaudhary, *Kamaiya System*, 36-37.

11. We should, at the same time, avoid making sweeping generalizations as to whether this added to *begaari* burdens, since there were conflicting implications. For example, while the number of public works increased, on one hand, the upgrading of irrigation canals using cement has lessened the need for repair, on the other. Moreover, the influx of public sector finances into villages contributed to undermining *begaari*, because it gave grounds for *Tharus* to make a case for implementing public works with government funds, in lieu of their free labor.

12. In Bardiya, all the MP (Member of Parliament) elections were won either by the NC and the CPN-UML during the 1990s. The two parties polled nearly equal votes, with the only exception that during the 1999 election, a breakaway faction of the CPN-UML had fielded a separate candidate, thereby splitting the votes. The RPP's MP candidates were mostly runners-up to the candidates from the two main parties, but usually fell far short of the poll of either the NC or the CPN-UML. There were two local elections during the 1990s. The CPN-UML attained a majority of the first DDC, and the CPN-UML continued to constitute a majority of the second Committee. In both the DDCs, the NC was the runner-up to the CPN-UML, followed by the RPP.

13. Norman Fairclough, *Discourse and Social Change* (Cambridge: Polity Press, 1992), 65.

14. Multi-party competition also played its part in heterogenizing discursive practices of district-level political leaders. The RPP's leaders, most of whom had assumed the posts of local elected leaders during the party-less *Panchayat* era, deplored the way multi-party competition divided *Tharu* communities along partisan interests and thus made it more and more difficult to organize *begaari*. The RPP's leaders usually asserted that "people's participation" should be conducted in the absence of party competition. Politicians affiliated with the RPP, just as those of the other major parties, were in favor of implementing river control based on the traditional corvée practice. However, local leaders of the RPP, unlike

those of the other parties, tended to glorify the *Panchayat* era when people had obeyed their leadership more "willingly" than they did the popularly elected leaders in the 1990s.

15. Excerpts from my interview with a former member of the DDC in October 2000.

16. Excerpts from my interview of February 2001.

17. Excerpts from my interview with the chair of one VDC in Bardiya in October 2000.

18. At the DIO in Bardiya, as of September 2000, there were three engineers (including the Chief), five overseers, one field assistant, and several other clerical and support staff. In the main text, "officials" refer to the engineers and the overseers.

19. Excerpts from my interview of January 2001.

20. Excerpts from my interview of February 2001.

21. The DIO provided materials such as G.I. (galvanized iron) wire and sandbags on an urgent basis to those areas unexpectedly struck by severe floods. For this purpose, the DOI at the center retained a stock of G.I. wire that was made available to districts in need of emergency works.

22. In 1999, the regular budget was used to construct a structure to protect an irrigation system controlled by activists of the NC. The irrigation group had sent a delegate to Kathmandu to get the NC government to instruct the DOI to deliver the project directly to the hands of the NC-led irrigation group. As explained in the next paragraph in the main text, this reflected increasing partisan considerations on the side of the central government.

23. Michel Foucault, "Two Lectures," in *Power/Knowledge: Selected Interviews and Other Writings 1972-1977 Michel Foucault*, ed. Colin Gordon (New York: Harvester Wheatsheaf, 1980), 99-100.

24. Foucault, "Two Lectures," 99-100.

# Chapter 5
# Village-level Unfolding of "Participatory" River Control (Majuwa)

This and the following chapters analyze the policy process of "participatory" river control at the village level. I focus on individual river control projects, because the policy under review took the form of discrete projects aimed at constructing river control structures. This chapter takes up a case of a village called Majuwa, where projects were undertaken in 1993, 1994, and 1996. Under the first and the last projects, the "participatory" river control policy was translated into action by drawing on the local corvée tradition in Western Nepal of *begaari*. The 1994 undertaking, on the contrary, was conducted through the use of a contractor.

Before examining the river control projects, the first section of this chapter provides a historical overview of how village politics evolved in the case study area. A key feature of local power dynamics in Majuwa was the historical domination of migrant *Pahadis* over native *Tharus*, who had long suffered the double burden of high rents and corvée obligations. At the same time, such a dichotomous view does not capture the entirety of village politics, in which the inequitable *Pahadi-Tharu* relations were constantly being readjusted, owing to the "duality of structure" which facilitated *Tharus'* resistance. The renegotiated nature of local dynamics also emanated from the vulnerability of "disciplinary" power upholding the age-old standard of *Pahadi* dominance. This had important implications for how the "participatory" river control policy unfolded in Majuwa where *Pahadi* villagers appropriated the policy to uphold the *begaari* tradition. Owing to the contested nature of local power dynamics, the modality of the river control projects was constantly remolded, often in favor of *Tharus*. This chapter thus indicates that a potential for the "emancipative" unfolding of the "participatory" river control policy was immanent in the daily interactions among policy "clients." This finding negates the need to resort to deliberate interventions, to correct the manipulative and exclusionary nature inherent in the policy process.

After the 1996 project, in the Bardiya district, there were few undertakings that followed the "participatory" river control policy. As stated in chapter 4, it is possible to make sense of the decline of the policy in the district, by analyzing how central political leaders came to belittle the "participatory" river control policy as a means of dispensing patronage. In contrast to such a conventional "descending" view focusing on "power holders", the move to do away with the policy can also be attributed to the intricate contestation over the policy direction among different actors in Bardiya, both at the district and the village levels. This chapter thus concludes by highlighting the importance of an "ascending" approach that heeds the micro-power struggles which brought about twists and turns in the local unfolding of the "participatory" river control directive, ultimately leading to the downward trend of the policy in Bardiya.

# Overview of Village Politics

The case to be taken up in this chapter is from Majuwa, of which real name is withheld to maintain confidentiality. The village corresponds to one administrative ward of a VDC (Village Development Committee).[1] The study area is of quadrangular shape, two sides of which are bordered by one river. After an exceptionally severe flood in 1951, the part of the river before the sharp bend started shifting its course towards the south, nearly one kilometer in width, thereby causing the northern side of the riverbank to be washed away. Before the 1950s, there had been a thick forest along the river to the north of the village. As a consequence of the bank erosion, part of the forest and some of its surrounding farmland were lost.

As of November 2000, there were a total of 138 households within Majuwa. Thirty-eight households were *Pahadi* families of hill origin, while the rest were households of *Tharus*, that is original settlers of the Tarai. *Tharus* therefore constituted a majority of the village population. Twenty-five of all the *Pahadi* households are of high caste, while the rest of *Pahadis* were either *Matwalis*[2] or of untouchable caste. The latter two groups of *Pahadis* hardly figured in the historical power contestation over *begaari*. First, since none of them worked as tenants or *kamaiyas*, nor used irrigation canals as cultivators, they were rarely obliged to participate in public works.[3] Second, although they were required to provide unpaid labor occasionally in the past, they were usually asked to "give a hand" not daily but only irregularly. This was also the case with the river control projects. Non-high-caste *Pahadis* did not feel resistant to the projects as much as *Tharus*, who had a historically deep-seated grudge against *begaari*. In this chapter, the term *Pahadis* denotes the twenty-five high-caste households.[4]

## Historical *Pahadi-Tharu* Rivalries

In line with the general picture of the Bardiya district, described in chapter 4, when *Pahadi* landlords of hill origin took over the village in the nineteenth century, they started exacting unpaid labor from *Tharus*. Before the land reform program in 1964, the entire area of the Majuwa was owned by two *Pahadi* landlords who also had properties in the surrounding villages. In those days, all the *Tharu* households worked as *kamaiyas* for the *Pahadi* landlords, except for the people in three houses who had managed to gain the status of tenants.[5] At the onset of the land reform program, the two landlords sold part of their land to

*Pahadi* owner-cultivators who had just arrived from the hills in search of new opportunities. Most landed *Pahadi* households migrated to the village during the 1960s. On the other hand, the majority of *Tharus* were not able to obtain land nor even to register their legally entitled tenancy rights because, according to many of the *Tharu* informants, *Tharus* had been kept in the dark about the land reform. Incised on the minds of *Tharus* was the conception that *Pahadis* manipulated governmental programs in collusion with government functionaries. The land distribution within the village continued to remain skewed in favor of *Pahadis*.

The influx of new immigrants of hill origin not only led to the continued reliance by a majority of *Tharus* on others' land, but also resulted in the loss of forests in and around the village. Many *Tharus* felt that they had been deprived of their resources which used to provide not only wood and grass for various purposes, but also opportunities for hunting animals. Another significant factor that contributed to deforestation was a resettlement scheme on the other side of the river, to develop a forest area into a market town. The government had originally planned to distribute land to a group of *Tharus*, who had come from different parts of the Bardiya district. However, they were later removed from the grantees' list, to be replaced by a group of *Pahadis* from a northern part of the Bardiya district. This incident was engraved on the minds of *Tharus*, as an episode that signified the motive of the *Pahadi*-dominated local functionaries, to keep *Tharus* dependent on others people's land and thereby to assure an ample supply of laborers for *Pahadi* landlords.

## Complexity of Village Power Dynamics

It is to be noted, at the same time, that such a polarized picture of ethnic rivalries as described above does not capture the entirety of the *Pahadi-Tharu* relations. Village-level power dynamics were not tantamount to a binary process of *Tharus'* resistance to the historical *Pahadi* dominance, and *Pahadis'* attempt to perpetuate it. On the contrary, the subjectivities of villagers were far from being neatly defined, and were instead multifaceted and fragmented. What facilitated the fluid nature of their identities were the inherent inability of discourse to accord fixed positions to social actors, and the resultant impossibility of prevailing societal norms, founded on certain discursive structures, to forge complete compliance.

*Pahadis* justified their historical dominance in the village, by depicting themselves as agents that had brought development into the backwater originally inhabited by *Tharus*. Such self-claimed righteousness as the "change agent" compelled *Pahadis* to face a dilemma over whether nor not to loosen their grip on the village because the assertion also required them to act as "benevolent benefactors." The following remark by the VDC chair, who described how he was coping with the diminishing "disciplinary" power of the *begaari* norm, attests to the need to work to uphold *Pahadis'* leadership position by projecting them as "trustees" who promoted the development of the "outmoded" village.

> "Nowadays it [the mobilization of labor contributions] is difficult. In those days, landlords scolded, and lots of work was done. Now, for example, we must make them understand that school must be constructed. School benefits them in this way. 'We are backward, and without school, we do not make progress. If the society does not progress, your son and grandson will remain like us.' . . . They must understand."[6]

The above type of reasoning of the *Pahadi* dominance was equivalent to a double-edged argument, serving both to uphold and to negate it. This is in line with the "open texturedness" of discourse, described in chapter 2, namely that discourse cannot impose order with finality, but is penetrated by competing forces. Accordingly, the historical norm of the *Pahadi* supremacy, articulated through such precarious and contingent assertion, was inherently vulnerable to other alternative interpretations, which would require *Pahadis* to relinquish their control.

Another factor behind the complex nature of *Pahadi-Tharu* relations was the advent of the multi-party system in 1990 that opened up more space for marginal groups to contest power along the lines not only of ethnic affiliations, but also of party, class, gender and other social and economic differences. As a result, the dichotomous picture of the *Pahadi-Tharu* rivalries does not grasp the entire spectrum of local politics in Majuwa, which instead could be explained better through the "structuration" perspective taken up in chapter 2. Unlike the "power as domination" view in which power is perceived to be absent, except when exercised, the "structuration" view considers it to be immanent in day-to-day social interactions, in which structure not only constrains but also enables human agency. In line with the "structuration" notion, power relations in Majuwa were renegotiated among villagers on a continual basis. In their ongoing power struggles, moreover, their structural locations in society were turned into "structural properties" that they draw on to challenge existing social order.

This "duality" of the structural inequality was exemplified by the politics surrounding corvée-based public works in Majuwa. With a view to standing in the way of the Community Party of Nepal-Unified Marxist-Leninist (CPN-UML) that occupied all the elected leaders' posts in Majuwa in the 1990s, activists from the Nepali Congress (NC) capitalized on the increasing discontent among *Tharus* with the inequality entailed in the *begaari* practice. In order to add fuel to their ill-feeling, NC activists also drew upon the perception commonly held among *Tharus* that *Pahadi* leaders embezzled part of the budgets in collusion with government officials, while making up for the misused funds with *Tharus'* free labor. Accordingly, there existed more and more openings for *Tharus*, of memberships of user committees for village public works, most of which had been occupied by *Pahadis*. The VDC chair[7] appointed an increasing number of *Tharus* in leadership positions, with a view to counteracting the NC's ploy to augment the preconception that *Pahadi* leaders were prone to corruption. This was the case with the river control user committee, to be reviewed in the following sections.[8]

In this way, the on-going party politics spilt over to the existing ethnic divide in such a manner as to allow *Tharus* to turn their inferior positions into their advantage. Moreover, in Majuwa, the "structural" disparity in land holdings not only subjugated disadvantaged *Tharus* to exploitative terms of employment, but also served as "structural properties" that they were able to draw upon to renegotiate existing social order. After the 1990 political change, in many areas of the country, including various localities in Bardiya, those who possessed little or no land came to resort to squatting on unused government land to demand that they should be allowed to register the plots they occupied, or else be provided with alternative land. This type of collective bargaining was usually supported by different political parties which competed with one another to dispense patronage to their respective support bases. These initiatives did not fundamentally alter the pattern of land ownership. At the same time, the success in obtaining land, albeit small in size, with recourse to political parties, inspired poor *Tharus* to further efforts to renegotiate their structural location in society, which they realized could be "structural properties" for challenging social order. This is attested to the "participatory" river control projects, to be taken up in the ensuing sections, in which *Tharus* ceaselessly challenged the inequitable *begaari* practice through both open, publicly declared confrontations, and covert, unobtrusive resistance.

The mainstream "power as domination" view, as described in chapter 2, presumes that powerful A acts in such a manner as to harm less powerful B's interests. In making sense of the increasingly intricate village power dynamics, however, it is more useful to adopt the "structuration" notion in which power is considered to be instantiated in daily social interactions, allowing social actors to respond to both the limitations and opportunities arising from the daily lives, and to ceaselessly renegotiate ongoing social order. Underlying this "duality" of village power dynamics was also the inherently fluid and inconsistent nature of villagers' subject positions, described above. The patron-client norm prevailing in Majuwa supposedly served to "discipline" villagers to embrace the historical *Pahadi* dominance, but at the same time, compelled *Pahadis* to loosen their control over *Tharus* as "benevolent benefactors." It is thereby imperative to avoid deducing the unfolding of village power contestation from various actors' structural locations in society, and instead to make careful empirical investigations, given the provisional and multifarious nature of individuals' identities. With the dynamic nature of village power struggles in mind, the following section dissects the policy process of "participatory" river control at the village level. For this purpose, since the policy was translated into action in the form of discrete projects, I examine three river control projects that were undertaken in 1993, 1994, and 1996.

# Village-level Contestation
# over "Participatory" River Control

As mentioned above, about fifty years ago, the part of the river before it reaches the sharp bend started shifting its course towards the village, and thereby caused riverside farmland and forest to be washed downstream. With a view to containing the erosion of the bank, seven riverbank protection structures were constructed in 1993, 1994, and 1996 (see Map 5.1). The 1993 and 1996 projects were undertaken through *begaari*, in line with the "participatory" river control policy, while the 1994 undertaking was entrusted to a contractor. As mentioned above, the historical *Pahadi* dominance and the concomitant *begaari* standard were ceaselessly readjusted, given the "dual" nature of the inequitable social order, which *Tharus* capitalized on to turn their inferior position to their advan-

**Map 5.1. River Control Projects in Majuwa**

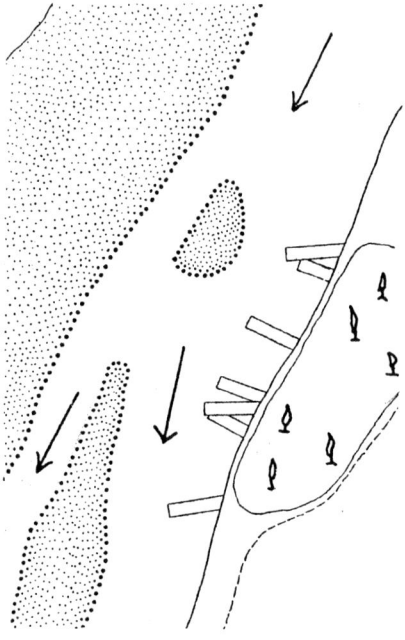

tage. Such "duality of structure" was also facilitated by the growingly complex nature of local politics, in which different facets of power struggles were interwoven especially after the 1990 political change. The *Pahadi* supremacy, in the first place, was inherently precarious and contingent in that it was articulated through the double-edged assertion requiring *Pahadis* to act as both "rulers" and "benevolent benefactors." This section examines how this intricate and contingent nature of the "disciplinary" power operating in Majuwa, where domination and resistance were interwoven, operated on the unfolding of the "participatory" river control policy in the village.

## Preparation for 1993 Project

Owing to the contested nature of the "disciplinary" power, which subjected the *Pahadi* dominance to ceaseless renegotiations among villagers, the mode of "people's participation" was already being debated constantly at the preparatory stage of the first river control project in 1993. The "participatory" river control project envisaged two categories of labor contributions, that is to transport boulders from the upper reaches of the river using bullock carts, and to work at the construction site in piling up boulders, holding them together with steel wire, and installing them in the water (see cover photo). With regards to the latter task, the VDC chair and the ward chair, who led discussions in village meetings,[9] had decided to propose that all the households, both *Pahadi* and *Tharu*, would provide unpaid laborers. Underlying this readjustment of the *begaari* standard was the "duality of structure" that intermingled domination and resistance in the daily flow of village social interactions. Given the ever-increasing resistance by *Tharus* against the historical *Pahadi* supremacy, the elected village leaders[10] had to work to reduce the inequality entailed in the *begaari* tradition in their attempt to revitalize and perpetuate it.

This alteration also emanated from the inherent vulnerability of reasoning behind the corvée tradition, which supposedly served to, but failed to "discipline" *Tharus* to acquiesce to its inequitable nature. The village leaders usually emphasized that river control was a "public" good, and that *Tharus* should act in the "public" interest as the residents of Majuwa. However, this logic also gave grounds for *Tharus* to demand that *Pahadis* would provide *begaari* on a par with *Tharus*. *Pahadis*' participation would be imperative if everybody's livelihood was to be saved with the limited resources. Riverbank erosion would eventually affect every resident, whether landlord, tenant, or landless laborer, unless the village was safeguarded from the potential risk of being washed away. This "surplus of meaning" attests to the "undecidability" of discursive structures, described in chapter 2. The *Pahadi* domination was inherently vulnerable to counter-discourses that argued for redressing its inequitable features.

As pointed out by Foucault, when and where "disciplinary" power is in operation, a "whole field of responses, reactions, results, and possible inventions may open up."[11] This was also the case with the 1993 project. For example, the provision to involve *Pahadis* side by side with *Tharus* backfired on landless *kamaiyas*. At the time of the 1993 project, forty-seven out of eighty-one *Tharu* households (fifty-eight percent[12]) earned their livelihoods working as *kamaiyas*. On the basis of the same logic that what was at stake was the "public" interest and that "everybody" should join hands, the elected village leaders had to oblige *kamaiyas* to depute one laborer from their own households. This would mean that although *kamaiyas* had historically had to represent their landlords' houses only, they would have to send an additional person on behalf of their own families. *Kamaiyas* expressed their dissent, on the grounds that they could not spare their time without pay, given their hand-to-mouth existences. The meeting eventually arrived at a compromise that a landless family with infants would have to participate only occasionally, not on a daily basis.

The new formula to require all the households, both *Pahadi* and *Tharu*, to provide labor at the project site also fueled *Tharus'* resistance to the other task of transporting boulders. It was originally proposed that, in line with the prevailing practice,[13] *Tharu* farmers would carry boulders without pay from the upper stream of the river. However, *Tharus* stood against the repetition of the conventional practice, drawing on the reasoning that carrying loads of boulders from upper reaches of the river would be far more demanding than other duties using carts. In response to this opposition from *Tharu* farmers, it was decided to provide remuneration but a reduced rate, by allocating part of the project budget. The project would provide 600 rupees, while the government rate was 800 rupees per *chatta* (approximately ten by ten by one inch).

In this way, from the start, different actors struggled with one another over the execution modality of the 1993 project. The river control undertaking represented not only resources to protect their villages from flooding, but also an additional domain in which village actors asserted or challenged existing social relations. The resultant tinkering of the corvée practice served to weaken the "disciplinary" power of the historical *Pahadi* dominance, that inclined villagers to take for granted the application of *begaari* to local public works. The following sections examine how the dynamic interactions observed at the onset of the 1993 project were carried over into subsequent activities.

## Construction of Structures in 1993

The 1993 project resulted in the installation of two bank protection structures, for which villagers labored for five days at the construction site. There existed a common sense of resentment towards *Pahadis* among *Tharus*, whether landlords, tenants, or *kamaiyas*. In the eyes of *Tharus*, the *Pahadi* participants in the 1993 project tended to put on the air of supervisors at the construction site, rather than working on a par with *Tharus*. At the same time, because it had been made clear that a defaulter would be fined a penalty or ostracized from the village, there existed no overt protest from *Tharus*. However, this does not mean that quiescence was pervasive. On the contrary, the modality of *begaari* continued to be renegotiated during the construction period, owing to the contested nature of the village power dynamics, explained in the preceding section. The diminishing "disciplinary" power of *begaari*, and the resultant remolding of the corvée rule during the 1993 project, took place within the context of ongoing village politics, in which *Tharus* managed to turn "structural constraints" into "structural properties", in response to opportunities arising from their daily lives. Under the 1993 project, this "duality of structure" manifested itself in the form of what Scott calls "infra-politics."[14]

According to Scott, subordinate groups engage in low-profile "infra-politics" to unobtrusively challenge the existing social order through covert, undisclosed, undeclared resistance. It does not only function as a building block for more open rebellion, but also constitutes a down-to-earth stratagem to minimize appropriation by dominant actors. This was also the case with the 1993 river control project in Majuwa, under which *Tharus* tried out subtle ways of shirking their tasks in various ways, while observing the overall *begaari* rules. *Tharu* peasants had a practical understanding of the difficulty of uprooting the corvée system, but tacitly knew that it would be possible to challenge it, albeit within limits, in a subtle and incremental manner. This "feel for the game" corresponds to what Bourdieu terms "habitus", or the system of "structured, structuring dispositions" inscribed in individuals' bodies, referred to in chapter 2.

For instance, children were often sent to the construction site to minimize their corvée obligations, because supervisors were not able to be as strict with them as they would be with adolescents or adults, and were therefore likely to give a lesser amount of tasks to children. According to the prevailing local corvée rule which linked one's labor requirements to the size of one's land, *Tharu* tenants who cultivated more than three *bighas*, or approximately two hectares, had to send two persons, while those who tilled less than three *bighas* deputed only one person. Out of the fifty-eight *Tharu* households that fell into the latter category, forty-three percent (twenty-five households) deputed children below the age of fifteen, at least for part of the construction period, and thirty-one percent (eighteen households) did not involve anybody more than fifteen years old at all, but only relied on minors. With regard to twenty-three households that sent two persons, thirty-five percent (eight households) left the entire *begaari* obligations to a pair of children less than fifteen years of age.

*Tharu* peasants also tried out the limits to which they could expand their discretion by defining for themselves the rules for some matters. For instance, *Pahadi* leaders attempted to exert control over *Tharus* concerning when to take a lunch break. When exhausted under the scorching midday sun, however, *Tharus* would start leaving their work in groups of twos and threes, but often many of them at the same time. Facing this tacit but well-coordinated strategy, village leaders could only let *Tharus* have their own way. Moreover, throughout the working day, some *Tharus* were also constantly waiting for a chance to rest under the shade of a tree, or even to sneak away from the construction site. Village leaders often had to play a cat-and-mouse game with *Tharus*, with a view to keeping them under strict discipline. This is illustrated by the following remarks by one elderly *Tharu*, whom many of his neighbors described as shrewd enough to shirk *begaari* obligations on many occasions.

"One *Pahadi* girl took pity on me [because I myself had to provide labor though I am very old], and said 'Please go home.' So I went home earlier than others. She was in charge of monitoring [taking attendance]. One person later came to my house, but I said 'I would not pay [the penalty] even if I get killed.' They could have killed my pigs or chicken if I had had any. But how could they kill a human being. So they went away."[15]

At the same time, this does not mean that all *Tharu* peasants did not shrink from adversity and maneuvered themselves out of work. As mentioned above, in view of the *kamaiyas'* fierce opposition, it was decided that a nuclear *kamaiya* family with infants would have to participate only occasionally. Not all of the landless eligible for the special treatment made use of the concession, and some instead sent one person or more everyday despite all hardships. Some *kamaiyas* feared that the powerful might rescind the rule, and might even exact penalties from them. There were also cases in which landlords abused this system, to instruct the wives of their *kamaiyas* to undertake *begaari* for a few days as per the concession, but forced them to work for the remaining days on behalf of the landlords. In this way, some of the destitute *Tharus* continued to be "disciplined" by the traditional asymmetrical *begaari* practice, and some landlords even responded with renewed repression, with a view to holding back the decline of the corvée tradition.

## Preparation of Contractor-based 1994 Project

In the following year, another river control project was undertaken in the village to construct two additional bank protection structures, but through a contractor. As mentioned already, *Pahadis* generally upheld the *begaari* tradition as the system through which they had introduced development into *Tharu* areas. Underlying the decision to avoid drawing on the corvée practice was the vulnerability of the *begaari* norm itself. The assertion about *Pahadis'* righteousness as "benign patrons" was bound to play into the hands of *Tharus* because it often compelled the village elected leaders to show leniency towards the plight of *Tharus* as "benevolent benefactors." This precarious nature of the *begaari* norm is attested to by the following remark by the VDC chair, who explained why the *begaari* practice was averted in 1994. The statement given below shows that the "disciplinary" corvée norm prompted the VDC chair to reconsider abiding by it. This resonates with the Foucauldian notion that power is not only oppressive but can also be productive since it facilitates actors to readjust their goals and strategies.

"People were not able to participate in the project because the budget came at the end of the year [that is around April and May]. At that time, villagers were already busy in their field. Therefore, the VDC agreed to use a contractor. Had we insisted on conducting [the project] through people's participation, the villagers might have agreed."[16]

The alteration in the project execution modality in 1994 also illustrates that the unfolding of the "participatory" river control policy in Majuwa cannot be simplistically relegated to a binary process of *Pahadis'* attempts to perpetuate, and *Tharus'* resistance to challenge the *begaari* norm. The village-level execution of the river control policy did not always progress in a consistent and "logical" manner, but entailed some "unexpected" turnarounds, owing to the unstable and often contradictory nature of people's subject positions. In Majuwa, *Pahadis* were far from being a coherent whole seeking to apply the *begaari* tradition to local river control, nor were *Tharus* monolithic in their resentment at *corvée*-based flood mitigation.

The fluid and inconsistent nature of villagers' identities calls for such an "ascending" analysis as described at the end of chapters 3 and 4, in contrast to a "descending" analysis that simplistically dichotomizes society into the powerful and the oppressed. The "ascending" approach draws attention to the feelings of ambivalence prevailing among the elected leaders in Majuwa who had to decide to drop the application of the corvée tradition in 1994. Their sense of dilemma had emanated from the decline of the *begaari* norm, and the concomitant prevalence of ill feeling widespread among *Tharu* peasants, many of whom had complained that they had to forgo their daily agricultural tasks, or other livelihood activities.

What added to the incoherent nature of villagers' subject positions was the increasing complexity of local politics that came to be played out at the dynamic intersections of party, ethnic, class, and gender struggles Party politics particularly played a crucial part in impelling the elected village leaders to opt out of the corvée practice. The VDC chair affiliated with the CPN-UML needed to clear himself of the suspicion that he had embezzled part of the project funds in 1993, a gossip spread by the NC in the opposition camp. In order to dispel doubt, the leader needed to give way to the use of a contractor, which would leave him no role in financial dealings. Whatever the case might be, the NC managed to capitalize on the perception prevalent among *Tharus* in Bardiya, referred to in chapter 4, that *Pahadis* historically preempted governmental resources using their ties with functionaries.

## Reinstatement of *Begaari* for 1996 Project

Just as *Pahadis* tended to harbor fragmented agenda, *Tharus* were also ambivalent about leaving the construction of river control structures to private builders. One assertion prevailing among political leaders in Bardiya was that that structures built by contractors were not durable because builders cut corners to maximize profits, and the District Irrigation Office (DIO) expected commission, not quality works, from its contractors. This view was also shared by many villagers, including *Tharus*.[17] *Tharu* peasants, whose livelihoods were also at stake, thereby held conflicting subject positions in that they wished to evade corvée and at the same time, resented contractors. The 1994 undertaking eventually turned out to provide a case for reinstating the *begaari* practice for the 1996 river control project because some of the structures built by the contractor in 1994 had been damaged by a flood in 1995. What made matters worse was that, towards the end of the construction, the contractor had fled the village without paying wages for local laborers. As a result, in village meetings, there was not any overt opposition to the reinstatement of *begaari* in 1996.

In this sense, both the village elected leaders and *Tharu* peasants, whose choices had been wavering between the use of *begaari* and that of a contractor, were commonly inclined to the former. This does not mean, at the same time, that *Tharu* peasants were resolved to abide by the corvée practice. On the contrary, the modality of *begaari* under the 1996 project continued to be renegotiated and reshaped. As already quoted several times in this book, Foucault points out that power dynamics open up a "whole field of responses, reactions, results, and possible inventions."[18] Owing to the productive and open-ended nature of power, the unfolding of the "participatory" river control policy did not stop going through twists and turns.

As a result of the continued renegotiations over the corvée rule, two major revisions were made to the modality of *begaari* under the 1996 project. First, it was decided to raise the remuneration to be paid to cart owners for the transportation of boulders, from 600 rupees per *chatta* (approximately ten by ten by one inch) to the government's rate of 800 rupees. Many cart owners had spent more than 600 rupees on repairs, since the bearings and wooden frames of the carts had often been damaged during the 1993 project. Moreover, in 1993, the *Tharu badghar* or leader in charge of organizing the transportation of materials, had found it difficult to gather a sufficient number of carts. Some reluctant cart owners had even resorted to avoiding the task, claiming that their carts or oxen were not in a working condition. The elected representatives therefore decided to fully pay the official wage.[19]

Second, as the chair of the user committee for the 1996 undertaking, the elected village leaders nominated the cart *badghar*. This marked a move away from the conventional mode of management for village public works, in which a *Pahadi*-led user committee was superimposed on the traditional *khel* system,[20] which *Pahadis* had historically used to exact unpaid labor from *Tharus*. This change was made because the 1993 project had been fraught with difficulties in mobilizing carts, which the *badghar* was expected to be in a better position to straighten out. The cart leader supposedly had an incentive to mediate successfully in disputes because his tenure as the *badghar* also hinged on how well he could execute decisions taken at village meetings. Another major factor behind the decision was the party politics that increasingly divided the *Pahadi* society, resulting in the fraud controversy in 1993, referred to in the preceding section. The growing divisions among *Pahadis* had prompted the selection of the *Tharu badghar* as the chair of the user committee, instead of awarding it to a *Pahadi*.

Under the 1996 project, three bank protection structures were built, on which villagers labored for six days on the construction site. Just as in the 1993 project, *Tharus* resorted to covert, unobtrusive resistance as described above, with a view to minimizing their workloads.[21] Moreover, there was an additional, insidious form of low-profile protest in that the number of *kamaiyas* was twenty-two at the time of the 1996 project, a sharp decrease from forty-seven three years before when the first project had been conducted. The 1993 project had added to the grievances that *kamaiyas* had been nursing against mistreatment by their landlords,[22] and had prompted more and more *kamaiyas* to negotiate with their masters to gain tenancy status.

## Post-1996 Riverbank Protection

In this way, the mode of *begaari* continued to be ceaselessly renegotiated throughout the 1993 and 1996 projects, which resulted in the continual readjustments to the corvée practice founded on the asymmetrical relations between *Pahadis* and *Tharus*. This corroborates the vulnerability of the *begaari* norm, described above, which not only exerted pressure on villagers to conform to the standard, but also facilitated them to reflect on and resist it. Such precariousness of the corvée standard was also attested to by the fact that the "participatory" river control undertakings had repercussions in the form of the post-1996 planting of trees and shrubs by the river, of which aim was to reinforce the riverbank protected by the 1993, 1994, and 1996 projects.

This was initiated by a mothers' group in Majuwa. Spurred on by the mushrooming growth of women's activities in surrounding villages in the 1990s, there had been a desire among group members to show their abilities to implement an idea that had never brought up nor discussed during the male-dominated decision making process. Women had not been present in the meetings, while only males gathered and discussed how to mobilize *begaari* contributions for the river control projects. This was in line with the tradition where only male heads of all the households were called together to decide how to run public works projects. This male-dominant nature of the corvée norm had provided grounds for the mothers' group to challenge and renegotiate the historical *begaari* standard through their riverbank planting program.

More importantly, this initiative had not only originated as a result of the female reaction to the male dominance, but had also been instigated by some NC activists, who had noticed that *Tharus* affiliated with the CPN-UML were cultivating part of the government land by the riverbank. In response, the cultivators, most of whom had little land of their own, and were threatened with potential eviction, lashed out by arguing that the destitute should not be deprived of their rights to subsist. In this way, the planting of trees and shrubs refracted with party, ethnic, and class conflicts in such a manner as to exacerbate the on-going village tensions. The unfolding of this program illustrates the infeasibility of executing public works projects without being entangled in village politics in which different facets of struggles were interwoven with one another. The following section considers how best to analyze the policy process of "participatory" river control, in the light of the difficulty of circumventing village tensions.

## Decline of "Participatory" River Control, and Village-level Micro-power

As explained in this chapter, the "participatory" river control policy served as a domain for various actors in Majuwa to enunciate or challenge the various inequalities entailed in the corvée-based river control efforts. As a result, the corvée rules for river control were in a state of flux, with the terms ceaselessly readjusted according to the particularities of the circumstances in which each project was undertaken. This was because, in the case study projects, the *begaari* standard not only exerted pressure on villagers to conform to the norm, but also served as a medium for local actors to contest existing social relations. This chapter thereby attests to the assertion of Foucault that "disciplinary" power not only subjugates individuals to constrained positions, but also facilitates them to ceaselessly contest the interpretations of pervasive norms.

The continual remolding of the *corvée* rules can be grasped within the context of ongoing village politics, in which T*harus* turned structural constraints into "structural properties", in line with Giddens' "structuration" view. *Tharu* peasants, whom the *begaari* obligations fell heavily on, continually engaged in what Scott terms "infra-politics" to work unobtrusively to manipulate the rules through undeclared, covert resistance. The "structural" constraint was thus turned into "structural properties", with which *Tharus* tried out subtle ways of shirking their *begaari* tasks, while avoiding openly rebelling against the *corvée* practice.

What added to this "duality of structure" of village power dynamics was the infeasibility of discourse to accord fixed positions to villagers, and the concomitant inability of prevailing societal norms, founded on certain discursive structures, to forge complete compliance. Accordingly, the *begaari* norm was inherently vulnerable to such counter-discourses as to condemn the injustice entailed in the corvée norm. Even those *Pahadis* who vehemently upheld the *begaari* tradition, therefore, did not seek to solicit *Tharus*' labor contributions at all times, but were constantly caught in a dilemma whether or not to reduce river control to the corvée-based construction of physical structures. Because villagers' identities were not fixed but were fluid, the execution of the policy in Majuwa did not always progress in a "logical" and purposive manner, but entailed several "unexpected" turnarounds, such as the decisions to use a contractor for the 1994 project, but to reinstate the *corvée* modality in the following year's project. This chapter accordingly illustrates that one should avoid reducing the policy process to a binary process of local elites' attempts to perpetuate, and *Tharus'* resistance to challenge the *begaari* norm.

Given such subtle and intricate workings of power, moreover, this chapter indicates that a potential for the "emancipative" unfolding of the "participatory" river control policy was immanent in the daily social interactions among policy "clients." In contrast to a pervasive assumption that a separate process of countermeasures should be launched, it was not a prerequisite, in the case study projects, for *Tharus* to engage in additional activities in order for them to exercise some leverage over the proceedings of the projects. Instead, the modification to the *begaari* mode evolved out of the daily struggles of *Tharu* peasants to constantly seek opportunities for renegotiating ongoing power dynamics, a finding that refutes the need to make deliberate interventions to ameliorate the manipulative and exclusionary nature of the policy process.

The immanency of "emancipative" potentials in the implementation of the policy on the ground can also be grasped in relation to the vulnerability, within Bardiya as a whole, of the "disciplinary" power of the *begaari* norm. As described in chapter 4, owing to the penetration into district leaders' discursive fields of such counter-discourses as to stress the importance of redressing the injustice embodied in the corvée tradition, the application of the *begaari* norm to river control undertakings was far from being a foregone conclusion. On the contrary, the project modality was constantly reflected on among policymakers in Bardiya who felt ambivalent about reducing "participatory" river control to the corvée-based construction of physical structures. The precariousness of the corvée standard at the district level, and the village-level renegotiations of the *begaari* modality in Majuwa provided fertile soils for each other. The former not only laid the ground for, but was also conditioned by the latter. Such synergy between the district-level dynamics and the grassroots struggles facilitated *Tharus* to engage in "cultural politics" over the manner in which how "participatory" river control was translated into action.

As stated at the end of chapter 4, in the Bardiya district, after the 1996 project, there were few river control projects that complied with the "participatory" river control policy. Drawing on a conventional "descending" analysis that considers power to be exercised by dominant actors over others, the policy decline can be attributed to the changed political contexts in Nepal. To bypass the CPN-UML that dominated many of the DDCs and VDCs, including those of Bardiya, the NC in power at the center had been exerting influence on the DOI to deliver river control resources directly to the party's local cadre, disregarding local elected representatives.

According to an alternative "ascending" perspective, the decline of the policy can be explained, in the light of the contestation over the definition of the policy among different actors, both at the district, and the village levels, which brought about twists and turns in the unfolding of the "participatory" river control policy in Bardiya. In this way, the "ascending" approach, by heeding the micro-power experienced at the village level, brings to light factors that would otherwise remain unnoticed by analysts who examine the policy process only abstractly from the presumable motives of high-level policymakers. The next chapter continues to undertake "ascending" analyses in another village in Bardiya, where *begaari*-based river control projects were initiated around the turn of the millennium, although the "participatory" policy had long faded away in Bardiya.

## Notes

1. As explained in chapter 3, the VDC was divided into nine electoral districts, in each of which a Ward Committee was formed. The Ward Committee was composed of one chair and four members. One of the four ward memberships was reserved for a female representative. The ward chair also served as a member of the VDC. The VDC chair, the vice-chair, and all the ward representatives were directly elected by residents.

2. Various ethnic groups in the hills and the Tarai were labeled as *Matwalis* or Alcohol Drinkers in the nineteenth century by the then ruling family of the country who intended to bring them under the Hindu hierarchy and thereby to proclaim their rule of the country. The two *Pahadi Matwali* households in the village were of ethnic minorities in the hills.

3. Non-high-caste *Pahadis* earned their livelihoods from a combination of livestock rearing, backyard farming and labor work in nearby towns. The *begaari* practice had come into being originally as a way of maintaining irrigation canals, as described in chapter 4, and had long continued to be applied only to those using the irrigation system.

4. Power contestation over *begaari* in Majuwa evolved along the line of historical "ethnic" rivalries between *Pahadis* and *Tharus*, rather than as "caste" struggles. The two were not mutually exclusive, however, since the former tension partly also emanated from the imposition of the *Matwali* caste on *Tharus* to bring them under the Hindu hierarchy. However, *Tharus* and other *Pahadi Matwalis* did not usually identify themselves as one group in the village

5. As explained in chapter 4, *kamaiyas* are agricultural laborers whose families were obliged, in return for a meager amount of grain, to undertake any tasks as demanded by their landlords. On the other hand, tenants used their own equipment and oxen, and were entitled to retain a certain portion of the yield.

6. Excerpts from my interview of October 2000.

7. The VDC chair, who was also himself a resident in Majuwa, figured in the management of public works within the village although it corresponds to one of the eight wards of his jurisdiction. During the same period, the elected ward chair was served by a *Tharu* leader, also from the CPN-UML.

8. The intensification of such local-level party politics mirrored that of party rivalries at the center. Nepal had as many as ten governments during the 1990s. The frequent government changes were caused, in many instances, when ideologically divergent parties

formed an alliance of convenience. The multi-party system thus degenerated into a "naked power struggle", to use the phrase found in Hachhethu's work (2000, 91), and the preoccupation of political parties was to achieve power. This fierce party competition compelled party leaders at the center to exert pressure on their sub-national cadres to strengthen their party support bases by fair means or foul. This caused party rivalries to prevail in many programs in the village, including the river control projects.

9. In this village, a user committee was usually formed for externally-supported projects, and the chair's post was usually filled by an activist affiliated with the NC. However, the chair from the NC usually did not extend full cooperation. The two elected leaders residing in the village, that is the VDC chair and the ward chair, therefore took the lead in the proceedings of village meetings, and for this purpose, usually put forward proposals for discussions. Before executing public works in the village, male heads of all the households were invited to village meetings, to discuss how to implement public works.

10. In this chapter, such expressions as "the elected representative" or "the village leaders" are used to refer to the VDC chair and the ward chair, unless otherwise specified. The two leaders, both of whom were affiliated with the CPN-UML, occupied the elected posts throughout the 1990s, and as indicated in the preceding footnote, they played major roles in executing *begaari*. The ward chair was a *Tharu* and admitted to me that he did not like to see his kin and neighbors experience hardships. At the same time, the *Tharu* leader never confronted the *begaari* norm in public, and on the contrary, publicly declared that he was a keen advocate of *begaari*-based public works.

11. Michel Foucault, "Afterword: The Subject and Power," in *Michel Foucault: Beyond Structuralism and Hermeneutics*, ed. Hubert L. Dreyfus and Paul Rabinow (Brighton: The Harvester Press, 1982), 220.

12. This and other statistical data concerning people's participation, presented in this chapter, are based on the questionnaire survey, referred to in chapter 1. As stated in the introductory chapter, I surveyed all the households in Majuwa.

13. For village public works, when it was necessary to carry materials, such as wood and sand, bullock cart owners (corresponding to tenants or independent owner-cultivators) took charge of transportation without pay. In the case of cart owners using *kamaiyas*, the latter were usually ordered to transport materials on behalf of their masters. The transportation therefore had to be borne by *Tharus*, because virtually all *Pahadis* relied on *Tharu* tenants or *kamaiyas*.

14. James C. Scott, *Domination and the Arts of Resistance: Hidden Transcripts* (New Haven: Yale University Press, 1990).

15. Excerpts from my interview of March 2001.

16. Excerpts from my interview of October 2001.

17. Villagers' sense of suspicion also emanated from their controversies with the DIO under the 1993 and 1994 undertakings. In the case of the 1993 project, a dispute arose concerning the DIO-specified design, according to which the structures were built to meet the riverbank at right angles. According to villagers' views, however, the bank protection structures should have been installed at oblique angles to the riverbank, in order not to be hit by flood flows square on. Engineers at the DIO did not heed this opinion, but simply told the claimants to obey the technicians. At the time of the 1994 project, therefore, various village actors, including the VDC chair and some *Tharu* peasants pleaded for public hearings concerning the DIO-specified design of the new structures. The DIO, however, did not heed this request.

18. Foucault, "Afterword," 220.

19. Still, it had been anticipated that there would remain reluctance especially among those who would have to make more than double the number of trips to finish one *chatta*, compared with those who had stronger oxen and carts. To ensure an ample supply of bullock carts, under the 1996 undertaking, cart owners were organized into three-person teams to hold them collectively responsible for transporting boulders. In addition, some *Pahadis* were also sent upstream of the river, to help *Tharus* load boulders.

20. As described in chapter 4, *khel* was a communal labor system that *Tharus* traditionally drew on, to organize collective activities, such as rituals and agricultural activities.

21. For example, out of sixty-four *Tharu* households obliged to depute one laborer, forty-two percent (twenty-seven households) delegated children below the age of fifteen, at least part of the construction period. As stated already, children were considered by *Tharus* as a way of shirking their *begaari* obligations because village leaders would not drive children hard. On the construction site, *Tharus* also tried to push the limits to which they could shirk their tasks, just as in the 1993 project, by taking a rest under trees, or even sneaking away from the workplace.

22. As stated earlier, under the 1993 project, each *kamaiya* household had been obliged to depute one person, in addition to acting on behalf of their landlords. Moreover, almost all of their landlords had not handed remuneration over to *kamaiyas* although they transported the boulders on behalf of their landowners.

# Chapter 6
# Village-level Unfolding of "Participatory" River Control (Pyauli)

In this chapter, another case study is presented of a village called Pyauli, where flood control structures were constructed in 1999, 2000, and 2001. This chapter begins by providing a historical background of village politics. A key feature of village power dynamics was the historical dominance of migrant *Pahadis* over native *Tharus* and other marginal groups, such as landless squatters called *sukumbasis*, and *dalits* who were members of the untouchable caste. Just as in Majuwa in the preceding chapter, the inequitable social order was constantly readjusted, owing to the "duality of structure" and the precariousness of the "disciplinary" power of the *Pahadi* supremacy. The renegotiated nature of local power dynamics had important implications for the village-level execution of the "participatory" river control policy. The policy process in Pyauli not only entailed constant renegotiations over whether and how to apply the corvée tradition called *begaari*, but also resulted in constant twists and turns, including the "unexpected" cancellation and reinstatement of *begaari*-based river control in 2000 and 2001. The continual readjustments made to the project modality indicate that the potential for the "emancipative" unfolding of policymaking was immanent in the implementation of the "participatory" river control policy on the ground.

As stated in chapters 4 and 5, in the Bardiya district, there were few projects that followed the "participatory" river control policy, towards the end of the 1990s, owing to the increasing tendency of the government to impose centrally directed projects. In order to make sense of the downward trend of the "participatory" river control policy, this chapter proposes to pay attention to ongoing micro-level power struggles, in which various actors constantly readjusted and modified their positions in response to the limitations and opportunities that emerged during the policy implementation. Such an "ascending" perspective attributes the policy decline to the contested nature of the policy implementation on the ground, as well as to the constant renegotiations of the overall policy direction among district-level actors, described in chapter 4. According to this "ascending" view, village-level actors capitalized on the fluid nature of the policy process, to cause the corvée-based modality of "participatory" river control to decline in the latter half of the 1990s.

# Overview of Village Politics

The second case study is from a VDC (Village Development Committee), posi-
tioned at the intersection of trunk roads leading to some major towns. The VDC
functions as the seat of several government offices, and a market center that
serves not only this but other neighboring VDCs. The river control projects under
review were conducted in Ward Five, hereinafter called Pyauli, which is located
at the northern end of the VDC.[1] During a flood in June 1999, more than fifteen
hectares of land had been washed downstream all at once in one *Tharu* settlement
within Ward Five. Rapid riverbank erosion had posed an imminent threat to the
market center, approximately four and a half kilometers away from the river
cutting site.

As of November 2000, there were a total of 556 households in Pyauli,
amounting to nearly five times the size of Majuwa taken up in the preceding
chapter (although both of the two study sites corresponded to the administrative
unit of a single ward). Unlike Majuwa where most of the residents were *Tharus*
on whom forced labor historically weighed, *Pahadi* households (309) constituted
a majority in this case study area. *Tharu* households (234)[2] accounted for less
than a half, and the remaining, smallest portion of the households (thirteen) were
those of *Dheshis*[3] who belonged to Hindu caste groups, originating from the
southern *Tarai* plain. Many *Pahadis* in this case study village settled as *sukumbasis* or landless squatters, which served to set at odds various groups among
*Pahadi* residents. This also contributed to the eventual emergence of the multiple
parties that were active in village politics. Four parties fielded candidates for the
1997 local election, unlike Majuwa in which only two parties were dominant. All
these factors rendered it relatively difficult to conduct a community labor project
in this area, as illustrated by the river control projects. Before proceeding to the
analysis of the village-level policy process of "participatory" river control, the
following two sections explain how village politics evolved in Pyauli.

## Historical *Pahadi-Tharu* Rivalries

In line with the overall picture of the Bardiya district described in chapter 4,
Payuli was originally inhabited by *Tharus* and a limited number of *Dheshis*, until
the Nepalese state took over their land during the nineteenth century, and entrusted new *Pahadi* landlords of hill origin, to bring the newly acquired land
under cultivation and to collect taxes. Before the 1960s when the land reform
program was undertaken, there were only two settlements inside the village. The
residents of the two localities were almost all *Tharus*, although the entire land was
under the ownership of the new *Pahadi* landlords. As was the case with other
areas of Bardiya, in those days virtually all the old settlers had to work as laborers
or tenants for the *Pahadis*, and were obliged to provide *begaari* or unpaid labor

for the maintenance and repair of physical infrastructure, such as irrigation canals, roads, and bridges. Unlike Majuwa, at the time of the 1964 land reform program, the *Pahadis* completely parted with their land, partly because, a few years before land reform, the irrigation canals had become obsolete, due to a drastic change in the course of the river. This provided opportunities for *Tharu* leaders from the two existing settlements, to obtain title to the land abandoned by the *Pahadi* landlords.

At about the same time, there was also an influx of new *Pahadi* migrants from the hills. Some *Pahadi* migrants had ties with political leaders and government bureaucrats, both of which had been dominated by *Pahadis*. According to *Tharu* informants, some *Pahadis* drew on their acquaintance with the legal process and their influence over local functionaries, and managed to evict some *Tharu* claimants from the land, by such means as deceiving them into signing away their land, or getting government officials to forge false ownership papers. As a result, land distribution eventually became highly skewed in favor of *Pahadis*, as attested to by the fact that there were few *Tharus* in the two old settlements who were able to make their living as owner-cultivators, without recourse to tenancy arrangements with other landlords, most of whom were *Pahadis*. As of November 2001, there were only four *Tharu* landowners in the two settlements who were able to afford to provide tenancy to others.

What made village power struggles more dynamic and complicated was the existence of abundant forests in the middle of the twentieth century, which covered most of what is currently the area of Pyauli. There was also a substantial amount of unregistered cultivated land on the forest fringes, which had previously been used by the landlords. Some of the earlier *Pahadi* migrants settled on such open spaces or forest areas as *sukumbasis* or landless squatters. In addition, because the river started flowing in another direction in the early 1960s, new land had become available for cultivation in and around the old river course. It is to be noted that the skewed distribution of land in Pyauli therefore did not mirror the ethnic divide as much as the previous case study area, since the incidence of landlessness was not limited to *Tharus*. Such polarized rivalries between *Tharus* and *Pahadis* as described in the preceding two paragraphs does not provide the entirety of local power dynamics.

## Complexity of Village Power Dynamics

Village politics were played out through complex power relations that were anchored around, not only ethnic, but also parry, class, gender, and other differences. This brought about an environment conducive for marginal actors to feel more daring to raise their grievances in public. One factor that facilitated *Tharus*' resistance to the historical *Pahadi* dominance was party politics. The Community Party of Nepal-Unified Marxist-Leninist (CPN-UML) was predominant in the village, and occupied all the elected posts in Pyauli in the 1990s. At the same time, the intensification of inter-party tensions in the village prompted other parties to decline to extend cooperation, and to stand in the way of the elected leaders.[4] This provided a fertile ground for *Tharus* to resist complying with the corvée tradition, in collaboration with leaders in the opposition camp, who were immersed in party politics.[5]

In this way, *Tharus* were able to take advantage of their historically disadvantaged positions to renegotiate their places in society, while *Pahadis* also allowed them to do so. Just as in Majuwa, one important implication was that it was not feasible to deduce the identities of villagers merely from the positions that respective actors occupied in relation to ethnic, class, and other differences. The subject positions of social actors were neither static nor neatly defined, but were unstable, multifaceted, and inconsistent. The conventional "power as domination" view, which regards power to be exercised by dominant actors over others, cannot grasp the fluid and fragmented nature of individuals' subjectivities. Just as is done in chapter 5, therefore, in analyzing power dynamics in Pyauli, it is crucial to draw on the "structuration" perspective that sheds light on how marginal actors turn their inferior positions to their advantage.

In Pyauli, it was not only *Tharus* but also other disadvantaged groups who, in line with the "structuraion" view of power, responded to opportunities arising from their daily social interactions. This was the case with *sukumbasis*. The "structural" disparity in landholdings not only reduced *sukumbasis* to being underlings of dominant actors, but also turned into "structural properties" that *sukumbasis* could draw on to renegotiate the preexisting social order. For example, with the intensification of party politics, their status as the landless provided *sukumbasis* with the leverage to demand and obtain help from political leaders.[6] Moreover, the prevailing discrimination against *sukumabasis* not only subjugated them to unfair treatments by other villagers, but also served as an instrument for them to refute the discrimination itself. Village leaders at times imposed forced labor on *sukumbasis* for village public works, by taking advantage of their anxiety about potential eviction from their localities. At the same time, such misfortune was often taken advantage of by *sukumbasis* as "structural properties" to prove their determination to stay in the village, and to seek to gain recognition as permanent dwellers.[7]

Even when dominant actors straightforwardly asserted their positions, therefore, ongoing entrenched disparities often turned out to serve as "structural properties" for marginal groups to resist and challenge entrenched social relations. This "duality" of the inequitable social order also derived from the inability of discursive structures to impose fixed meanings, and the resultant impossibility of prevailing societal norms, founded on certain discursive structures, to forge complete compliance. *Pahadis* often justified their historical dominance by describing *Tharus* as the underclass who should feel grateful for the initiatives of the newcomers of hill origin for various development activities. This type of reasoning entailed contradictory discursive elements, projecting *Tharus* as "underlings" who should bear corvée obligations, while depicting *Pahadis* as "benevolent benefactors" who gave a hand to *Tharus*. Accordingly, in Pyauli, *Pahadis* often had to commit themselves to provide labor for public works projects side by side with *Tharus*. At the same time, because most *Pahadis* usually shirked from taking part, and even those who were present at the sites usually assumed supervisory roles, the above argumentation that commended *Pahadis* as "benign patrons" proved to negate their leadership. It also provided grounds for *Tharus* to occasionally refuse to provide corvée.[8]

In Pyaluli, therefore, the various "disciplinary" norms, upheld by dominant actors with a view to subjugating marginal players, were constantly contested and reflected upon. This was exemplified by *Pahdis*' assertion of their dominance over *Tharus*, villagers' rationale of holding *sukumbasis* in contempt, or men's negation of women's initiatives for social activities. Such intricate local power dynamics cannot be grasped only through the "power as domination" view that regards power to be exercised by the privileged to dominate the underprivileged. It is necessary to draw on the "structuration" perspective, according to which domination and resistance cannot be neatly disentangled but are intermingled in daily social interactions. It is therefore imperative to avoid deducing the unfolding of village power contestation from various actors' structural locations in society, and instead to make careful empirical investigations into the provisional and multifarious nature of individuals' identities. In the light of such subtle and dynamic nature of power relations, the following section examines the policy process of "participatory" river control, with the focus on three projects executed in Pyauli in 1999, 2000, and 2001.

# Village-level Contestation
# over "Participatory" River Control

As mentioned above, a flood in August 1999 caused more than fifteen hectares of land to wash downstream in Pyauli. The erosion of the bank posed an imminent danger to the market center that also served as the seat of various government offices, approximately four and a half kilometers to the south of the river cutting site. Should the river swell next time, floodwater could potentially be diverted towards the center of the VDC. As stated in chapters 4 and 5, the "participatory" river control policy came to be ignored by the central government during the latter half of the 1990s. In the case of the 1999 project, however, in the absence of prior budget allocation, it was necessary to mobilize "people's participation" to deal with the rapid riverbank erosion. The VDC consequently requested the District Irrigation Office (DIO) in Bardiya for emergency support for such materials as iron wire and sandbags, and solicited labor contributions of villagers to construct bank protection structures on an urgent basis (see Map 6.1). Though the

1999 project was followed by two undertakings in 2000 and 2001, the two ensuing projects were not conducted through the mobilization of labor contributions, owing to the decreasing tenability of the *begaari* standard. The following section analyzes how the decline of the corvée tradition was catalyzed during the 1999 project.

## Untenability of *Begaari* under 1999 Project

The *begaari* practice that fell inequitably on *Tharus* was upheld often on the grounds that *Pahadis* guided *Tharus* to development. At the same time, as described above, this assertion compelled *Pahadis* to face the dilemma, as "benevolent benefactors", of having to work to reduce the inequality entailed in the tradition, while putting *begaari* into action. This double-edged nature of the "disciplinary" norm also facilitated *Tharus* to turn their inferior position into "structural properties" that they drew on to challenge the corvée practice. In the light of the precariousness of the *begaari* norm, during the preparation of the 1999 project, the VDC decided to mobilize labor contributions from outside Pyauli (which corresponds to Ward Five). A request was sent out to other wards of the

**Map 6.1. River Control Projects in Pyauli**

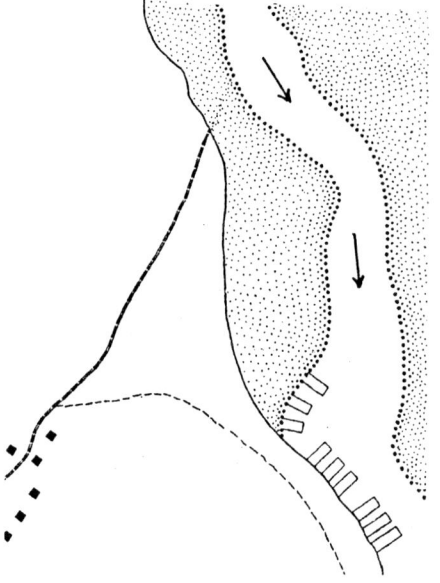

VDC, government offices located in the market center, and schools within the VDC, to extend assistance in constructing the structures. Village leaders justified this move on the grounds that the project was to safeguard wider areas beyond Ward Five, in the face of the imminent threat that the entire VDC might be swept away.

This logic served as what Foucault terms a "political technology" to divert attention away from the historical inequality entailed in *begaari* by highlighting the fact that what was at stake was the "public" interest. This type of reasoning served to highlight that river control was a public good beneficial for everybody regardless of their organizational affiliations, ethnicity, or other differences. At the same time, when it failed to lead to broad-based "public" involvement, it served as the basis of a counter-argument against the application of the *begaari* practice to local river control. Owing to this precarious and contingent nature of the assertion, what was supposedly a praiseworthy, collaborative effort turned out to defeat its purpose, in the face of the growingly dynamic and intricate village politics, as explained in the following sections.

*User Committee Afflicted with Party Politics*

Before the start of the 1999 project, it was decided, in order to mobilize the residents of Pyauli, to establish a user committee headed by the ward chair affiliated with the CPN-UML. The committee drew members not only from other elected ward members, all belonging to either the CPN-UML or its breakaway party CPN-ML,[9] but also from other non-elected leaders, in such a manner as to represent the two other main parties, that is the NC, and the RPP. The UC memberships consisted of ten members, three of whom were from the CPN-UML, three from the CPN-ML, two from the NC, and two from the RPP. However, the strategy to create an all-party user committee backfired. Each member was expected to mobilize people in his or her area of residence. However, most of the UC members, especially those from the NC and the RPP, only half-heartedly fulfilled their duties to mobilize residents in their respective areas.[10]

At the same time, committee members did not necessarily neglect giving clear-cut instructions to every resident. They were prone to be firm with disadvantaged segments of the community, such as *Tharus*, *sukumbasis*, and *dalits*, and thus gave these marginal groups detailed instructions as to the date and time of their participation, and even threatened punishment. Almost all the user committee members (eight out of ten) were high-caste *Pahadis*. The high-handed approach to the underprivileged groups reflected the disposition of *Pahadis* to tacitly attempt to shift the responsibility for labor contributions onto marginal groups, with a view to upholding the ongoing social order. At the same time, in line with the "duality of structure", the predicament facing *Tharus*, *sukumbasis*, and *dalits* also served as "structural properties" that they could draw upon to renegotiate the inequitable practice.

*Disparity in Participation along Ethnic Lines*

Under the 1999 project, most residents in Pyauli failed to participate in an organized manner. This was particularly the case with *Pahadis*, with the exception of some *sukumbasis* and *dalits*, to be taken up in the following section. Unlike *Tharus*, *Pahadi* participants did not organize a group in each settlement.[11] *Pahadis* instead turned up in small groups with their friends and neighbors. Moreover, each *Pahadi* group arrived at different times, and not all stayed for a long time. On the other hand, *Tharus* usually came to the project site together with other residents of their respective settlements. Almost all *Tharus* spent a full day there from morning to evening. There were a total of five *Tharu* settlements in the village, one of which was located at the land cutting site and was directly affected by riverbank erosion. The problem was not of immediate concern to the other four *Tharu* settlements. However, even *Tharus* from the four other areas took part in the project in a relatively organized way compared with *Pahadis*.

This does not mean that *Tharus* simply attended to the project without taking note of the inaction of *Pahadis*. *Tharus* implicitly knew that they could not abruptly uproot historical *Pahadi* dominance, and that they had to participate in the project in an organized manner as the underlings of *Pahadis*. At the same time, their ongoing social interactions with other villagers had accorded *Tharus* a practical understanding of the possibilities to challenge the inequitable practice unobtrusively through low-profile resistance. This "habitus" led some *Tharu* leaders to resort to applying lax rules in mobilizing their residents. In one community, for example, a leader designated three days, instead of one particular day, and the residents were able to attend on whichever date among the three suited them. In yet another locality, the *badghar* or the community leader did not mobilize all the residents, which he would normally do for village collective actions. The *badghar* justified this by referring to the manner in which one *Pahadi* leader had abruptly give instruction without prior notice or consultation.

> "We did not get information beforehand. One day Yam Bahadur [the name of the *Pahadi* user committee member in charge of this area] came here and told me that we must go the next day. How can we go if we hear only rumors [that I might have to organize villagers] without proper communications?"[12]

This statement highlights how the very high-handedness of *Pahadis*, supposedly intended to subjugate *Tharus*, provided grounds for them to ease up on their labor contributions. In this way, the ongoing entrenched inequality played into the hands of the *Tharu* leaders who capitalized on the "duality of structure" to ameliorate their historical inequality, albeit incrementally. The relaxation of *Tharus'* *begaari* practice, it is to be noted, cannot solely be attributed to such "structuration", but was also a reflection of the intrinsic precariousness of the corvée norm, already described in this chapter. The prevailing standard of the *Pahadi* superiority was based on their self-claimed status as "benefactors" of "backward" communities, which inherently required *Pahadis* to relax their control when *Tharus* found themselves in a dire predicament.

*Intersection with Sukumbasi Politics*

The unsystematic turnout of *Pahadis* was not observed universally, as stated above. Two *Pahadi sukumbasi* or squatter areas participated in the project in a relatively organized manner. One of these was the *dalit* settlement, referred to earlier in this chapter in relation to the path construction project.[13] The user committee member in charge of this area had told the *dalit sukumbasis* to participate for three consecutive days, and had even hinted that there would be a potential penalty for a defaulter. On the other hand, their non-*sukumbasi* neighbors were asked to go to the site for three days, whenever they had time. However, the *dalit sukumbasis* did not blindly follow the instructions. Their daily struggles with village leaders had accorded the *dalits* the practical sense that they would remain immune from punishment as long as they did not dare to publicly challenge the inequitable allocation of labor obligations. The following remark illustrates such a practical "feel for the game" regarding the *begaari* practice, with which villagers did not necessarily follow the instructions strictly.

> "We were told to start work at eight in the morning, but reached there at about ten o'clock. Because we live far [from the project site], we had our meal and left. Some [of us] left before four o'clock, because leaders knew that *sukumbasis* were busy [making ends meet]. But we were very willing to go. We thought 'if we do not go, we will be blamed', so we went together on the same day."[14]

The last sentence of the above quotation also indicates that, in the eyes of the *dalit sukumabasis*, the labor contributions for the river control project also presented an opportunity to gain recognition as fully-fledged residents in the village. Their landed neighbors considered that the squatters were not permanent residents and did not merit participation in various village activities. This reasoning played into the hands of the *sukumbasis* who showed their willingness to stay in Pyauli by making labor contributions to the river control project. In line with the conception of the "surplus of meaning" that denotes the vulnerability of discursive practices to counter-argumentation, the neighbors' sense of discrimination in itself served to allay ill feelings against *sukumbasis*.

Such "duality of structure", in which structure both constrains as well as facilitates one's endeavors to renegotiate power dynamics, was also observed in non-*Pahadi sukumbasi* settlements, including the *Dheshi sukumbasi* settlement. The *Dheshis* were liable to pressure from *Pahadi* leaders, upon whom they relied for protection against evictions, and upon whom many of them also depended for their livelihoods. They were therefore coaxed or coerced into providing labor at the project site. At the same time, the project not only posed difficulties for the landless, but also constituted a chance to display their allegiance to their protectors. This does not mean that the *Dheshis* were reconciled to their fate, or were only seeking to obtain recognition as residents of the village. On the contrary, just as the *dalit sukumbasis*, they tried out subtle ways of shirking their tasks by sneaking away from the project site while they were not observed, and by reducing the workloads by carrying smaller boulders. The *Dheshis* had a practical understanding of the difficulty of uprooting the inequitable social order, but tacitly knew that it would be possible to challenge it, albeit within limits, in such a discreet manner.

*Intersection with Gender Politics*

The analysis thus far made of the 1999 river control project amply illustrates the infeasibility of putting *begaari* into action without being entangled in power struggles, as village politics were played out through increasingly complex and dynamic social relations. This was also the case with women's forest groups that the VDC mobilized in the light of the overall thrust of the project to call on various stakeholders. The questionnaire survey conducted during the fieldwork[15] shows that males accounted for thirty-one percent (fifteen persons) of all the participants representing the forest groups (a sample of forty-eight persons). Women did not necessarily deplore the low level of female representation, but there was generally a sense of ambivalence. While hoping to represent their groups at the construction site, it was also of great relief, at the same time, to be exempted from the hard assignment under the scorching sun. The following statement by a member of one forest group shows how the male-dominance constituted the "duality of structure" that provided not only a constraint but also an opportunity for women to assert their positions in the river control project.

> "Women were not able to do hard work, such as working inside the river, or carrying heavy things. . . . So men also helped us. When women have small children, they had to prepare a meal and send [them] to school. We needed to help each other, and when women were busy, men had to go. If women do not do household tasks, men will be in trouble. Who cook meals and wash dishes? Who look after buffaloes? Before, we used to work as told by men. Now, it is like that, but it is slightly different. Men also work as told by women. If men do not listen to us, then we do not listen to them, either."[16]

Although men often dictated how the women's groups took part in the project, the very disadvantaged position of women also helped the female groups in that it excused its members for their absence. Moreover, even some of those who attended the project capitalized on the paternalistic attitude prevailing among men to leave the site early on the grounds that they were busy with their household work, or they were not feeling well, drawing on the myth that women were prone to get sick. In this way, women subtly and unobtrusively renegotiated power relations by virtue of their underprivileged positions that also served as "structural properties."

It is to be noted, at the same time, that the manner in which the forest groups participated in the project was also influenced by other dimensions of power struggles.[17] For example, the leaders of one group, established in a stronghold settlement of the NC party, did not even appeal to their members to take part in the project. In another group fraught by the division between the *Pahadi* and *Tharu* members, the former refused to work side by side, and thereby left the labor obligations to the latter. It is crucial to avoid categorizing these episodes neatly into party, ethnic, or whichever category of village politics.[18] An important point is that different facets of village politics were interwoven to forge multi-farious and often inconsistent identities of social actors. Women were not simply maneuvering themselves into renegotiating their social positions vis-à-vis men. As pointed out earlier in this chapter, various aspects of village politics were interwoven to bring about intricate and dynamic social relations.

*Resistance by the Directly-Affected Area*

As discussed above, the project's attempts to reactivate the corvée tradition backfired, such as the intimidation of the disadvantaged, the establishment of the all-party user committee, and the mobilization of female groups. This was also the case with the VDC's initiative to enlist help from outside Pyauli. As stated already, the VDC requested other wards, government offices located within its jurisdiction, and schools within the VDC, to provide a hand in constructing the bank protection structures. However, these groups from outside Pyauli generally did not make much contribution to the project.[19] On the other hand, since it was their settlement that was most imminently under threat, the residents at the construction site continued to participate on a rotational basis throughout the entire life of the project that lasted for twenty-two days. Even when those from within or outside Pyauli had stopped coming to the site, the inhabitants had to carry on the work, to ensure that the project was brought to completion.

In the eyes of the residents in the directly affected area, the river control project mirrored the conventional *begaari* practice, which had historically weighed heavily upon *Tharus*,[20] since most of government officials, school teachers, or ward chairs who had supposedly come to their aid were *Pahadis*. This type of reasoning, pointing to *Pahadis'* lack of commitment to *Tharus'* welfare, reveals the precariousness of the *begaari* tradition. The corvée norm, upheld by depicting *Pahadis* as "benevolent benefactors", was inevitably vulnerable to counter-discourses, since it contained the contradictory elements of relegating *Tharus* to being the underlings of *Pahadis*. Given the untenability of the argument for the corvée tradition, therefore, the historical norm of the *Pahadi* dominance was of a double-edged nature, serving both to uphold and to negate it.

What added to the frustration of the residents in the directly affected area was that, toward the end of the project, it was becoming more and more difficult to pile up sandbags in the river, since the water level had been rising day after day at the onset of the monsoon. Villagers who worked in the river had therefore been complaining of their plight, had appealed to the ward chair and the VDC chair, and successfully obtained funding to compensate themselves for their painstaking tasks in the water. Given the precariousness of the age-old *Pahadi* rule that could not impose social order with finality, the ward chair and the VDC chair had to agree to the demand, especially when the local *Tharu* society was putting their finger on the historical inequality that they had been attempting to divert attention away from.

Moreover, the decision to start paying those laboring in the river also prompted others working on the riverside to demand remuneration. For example, those *sukumbasis* who were providing a helping hand on the riverbank argued that they could not work for days without pay given their hand-to-mouth way of life. Women similarly made out a case for payment on the grounds that local leaders should acknowledge their contributions. Moreover, opposition parties were also inciting the workers to claim payment, with a view to standing in the way of the CPN-UML. The disputes about wages showed that *begaari* could no longer be put into action without being dragged into multifarious power contestation along the lines of party, ethnicity, class and gender. This left the ward chair and the VDC chair with no option but to agree to pay, towards the end of the project, those working on the riverside also.

## Use of Contractor for 2000 Project

The 1999 project was followed by two more river control undertakings in the two ensuing years. In 2000, additional spurs were constructed by a contractor. The Department of Irrigation (DOI) in Kathmandu had specified that its district-level river control budget be used for Pyauli using a contractor, in lieu of a blanket authorization as per the "participatory" river control policy. This was in line with the increasing propensities of the center, since the end of the 1990s, to impose predetermined projects on the District Irrigation Office (DIO), referred to in chapters 4 and 5. Owing to growing rivalries between the political parties in power at the center and those dominant in the district during the latter half of the 1990s, the "participatory" river control policy decreased its value, in the eyes of national politicians, as a means of distributing patronage.

As mentioned in chapter 4, the decline of the policy also manifested itself in Bardiya, where there were few projects that complied with the "participatory" policy, after the 1996 project reviewed in the preceding chapter. As stated above, the 1999 project in Pyauli was undertaken through "people's participation" on an exceptional basis, because of the urgent need to construct structures without prior budget allocation. Just as in most other districts, the CPN-UML occupied the bulk of the seats in the DDC/VDCs in Bardiya after the 1997 local election. The NC government in Kathmandu had been imposing centrally directed projects to Bardiya to deprive the elected politicians from the CPN-UML from taking credit for mobilizing labor contributions.

At the same time, such a "descending" perspective, paying attention to the strategies employed by "powerful" actors, offers a one-sided view of the policy process. It is also imperative to adopt an "ascending" approach that takes into account the subtlety of the micro-power experienced at the village level, in making sense of the unfolding of the policy process. This is because, as stated in the preceding sections, village power relations were ceaselessly renegotiated by various actors who constantly readjusted their standpoints in accordance with the limitations and opportunities that arose in their daily interactions. At the time of the preparations for the 2000 project, the ward chair and the VDC chair, who had been the driving force behind the corvée-based 1999 project, were not proposing outright the use of labor contributions. This ambivalent stance of the two leaders emanated from the "open texturedness" of the historical corvée norm that contained contradictory discursive elements of hailing *Pahadis* as "benevolent patrons" and of allowing them, at the same time, to exact unpaid labor from *Tharus*. Since, as described in the preceding section, *Tharus* in the directly affected area had to take on the bulk of physical work, it was necessary for the elected leader, as a "benefactor" of the village, to show leniency towards the plight of the *Tharu* residents. This dilemma of the VDC char is captured by the following statement.

> "The government gave budget this year, because we had done a lot of work without budget last year. We were also planning to implement through people's participation, but a flood came early, and it turned out to be impossible. Villagers were busy in their fields, but the contractor worked even in the rainy season to make money. . . . If people say 'we cannot [provide unpaid labor]', then, we have to give the money to a contractor."[21]

*Tharus* in the area directly affected felt relieved that they would not have to contribute labor, but at the same time, were left with a sense of ambivalence. This was because a conception prevailing in Bardiya that the DIO and contractors put commissions and profits before quality. From the viewpoint of a majority of villagers, especially the *Tharus* residing near the riverbank, this assertion was confirmed by the dispute about the design that they had had with district engineers during the 1999 project, in relation to the validity of the angle and the location of the structures. Villagers had insisted that structures should be built at some distance up the river, not at the river cutting area, and had also opined that diagonally placed structures would be more durable than the DIO's perpendicular design. Because some of the structures from the preceding year had shown signs that they might collapse in due course, The DIO was eventually compelled in 2000 to agree to construct structures diagonally in accordance with the villagers' insistence, although district engineers did not compromise the location.

## Failed Reinstatement of "People's Participation" in 2001

In the ensuing year 2001, the DOI planned for yet another river control undertaking for Pyauli, and initially proposed to hire a contractor. The 2000 undertaking turned out to offer a solid case for reinstating *begaari* for the 2001 project, since some of the structures built by the contractor had been washed away, thereby causing further riverbank erosion. The ward chair and the VDC chair, in deciding to undertake corvée-based river control, took into consideration a general feeling among the *Tharus* residing at the project site, whose livelihoods were also at stake, and who generally felt they themselves should handle the project in 2001 and gain more control over the construction, to ensure the durability of new structures. The disputes that the residents had had with the DIO under the 1999 and 2000 projects, concerning the design of the structures, brought home to the *Tharu* residents the notion prevailing in Bardiya that the DIO and contractors put profits before quality.

This "unexpected" turnaround, which ostensibly ran counter to the *Tharus'* resistance to the use of *begaari* under the 1999 project, attested to the "social antagonisms" described in chapter 2, that is the ultimate contingency of one's subject positions, given the precariousness of one's discursive practices. The *Tharus* did not necessarily resent the corvée practice at all times. On the contrary, they were prompted to uphold *begaari*-based public works when the modality proved more conducive to the protection of their livelihoods than did the use of contractors. In order to make sense of this seemingly "unforeseen" unfolding of events, it is imperative to adopt an "ascending" approach to delve into what Foucault terms "antagonisms of strategies"[22] and the resultantly subtle and con-tradictory nature of people's identities. One should start by examining the mi-cro-power experienced by different individuals and groups, and then try to reveal a more general, overarching power dynamics.[23]

The ward chair also faced "social antagonism", that is the dilemma of having to relax the *begaari* practice to move forward the corvée-based public works. In view of the various difficulties experienced under the 1999 project, the ward chair proposed a revised plan for mobilizing labor contributions. First, it was originally conceived that participation would be sought only from within Pyauli, consid-ering the failure in 1999 when seeking help from outside the village. Second, given the difficulty of getting residents to contribute free labor for an extended period, as revealed two years before, it was proposed to reward participants with a certain percentage of the market wage rate. Third, in the light of the infeasibility to mobilize residents in an ordered manner, a new user committee was formed with fifteen members, larger than the 1999 committee. The new committee fol-lowed the principle adopted two years before of drawing members from different political parties.

The 2001 project did not start for a long time, however, because the revamped user committee was not able to shape a work plan. The newly selected members did not openly object to the overall direction, but hardly took the initiative to lay the groundwork, such as sounding out their neighbors on new rules. As time was pressing, it was then decided to solicit labor contributions only from the directly affected *Tharu* settlement. However, the local residents were generally displeased with the course of the events, in parallel with the 1999 project under which the people in the locality had to take on most of the burden. Many *Tharus* conse-quently skipped their labor contributions, thus causing the 2001 undertaking to be suspended after a while. In this way, the idea of subsidized *begaari* was also dragged into local politics. This brought about difficulties for leaders to move forward public works with "people's participation" and eventually compelled the DIO, with a view to bringing the project to completion, to reverse the project modality back to the original plan to use a contractor.

## Decline of "Participatory" River Control, and Village-level Micro-power

In Payuli, the *begaari*-based river control projects were no longer executed without being entangled in intricate power struggles, where various villagers ceaselessly renegotiated their social standing. Underlying the contested process of the project implementation was the "duality of structure" which facilitated *Tharus* and other marginal groups to turn their inferior positions into "structural properties" to challenge the corvée practice. Under the 1999 project, some user committee members attempted to be firm with marginal groups, such as *Tharus*, *sukumbasis*, and *dalits*, often by threatening punishment. While those disadvantaged groups avoided openly resisting the coercion, they drew upon their predicament to ease up on labor contributions. Their daily interactions with other actors in the village, in which the inequitable social order was constantly being renegotiated, accorded them a practical understanding of the possibilities of slackening off their labor contributions in a low-profile way. On the other hand, *Tharus* in the directly affected settlement, who continued working to bring the 1999 project to completion, resorted to more open resistance by demanding remuneration for their contributions. The inadequate participation by those from outside the project area accorded them the grounds to publicly resist the corvée practice that weighed heavily on them.

The "duality of structure" was furthered by the vulnerability of the prevailing norm of the unequal social order and the resultant contingency of villagers' subject positions. The inequitable *begaari* standard was often upheld on the grounds that *Pahadis* had brought development into the backwater originally inhabited by *Tharus*. Such high-handed argumentation reflected the ongoing social order, in which *Pahadi* elites had historically shifted *begaari* burden onto disadvantaged segments of the population. At the same time, the reasoning subjected village elites to the "social antagonism" of having to work, as "benevolent benefactors", to relax the corvée practice. Given this double-edged nature of discourses, the hierarchical social order, founded on various articulatory practices, was bound, not only to constrain human actions, but also to enable individuals to seek opportunities to renegotiate ongoing social relations. For example, even when village leaders forcefully exacted *begaari* from *sukumbasis* by capitalizing on the latter's lack of security against eviction, the *sukumbasis*' disadvantaged position served as a means to assert their rights as permanent residents. In this way, social domination often played into the hands of deprived segments of the village populace.

The implications of such subtle power dynamics were twofold. First, it is not feasible to reduce the grassroots-level unfolding of the "participatory" river control policy, to a binary process of assimilation with, and resistance to the *beggari* norm enunciated by local elites. On the contrary, given the "dual" nature of the inequitable social order, and the resultant fluidity of villagers' subjectivities, the village-level policy process of "participatory" river control did not unfold in a linear and "logical" fashion, It entailed twists and turns, including such "unexpected" turns of events as the decision in 2000 to use a contractor, and the failed attempt to reinstate *begaari*-based river control in the ensuing year. More importantly, the continual readjustments made to the project modality in Payuli illustrate that the potential for the "emancipative" unfolding of the "participatory" river control policy was immanent in the implementation of policy on the ground. It is imperative to avoid presuming the necessity of initiating a distinct set of interventions to counteract the manipulative and exclusionary nature of the policy process, contrary to the prevailing assumption about the need to make deliberate interventions.

The analysis of the policy process of "participatory" river control in Pyauli, as in the case of Majuwa, indicates the importance of an "ascending" perspective that takes into account the subtle micro-power struggles experienced by policy "clients", and their resultant sense of "social antagonism." The "ascending" analysis of the Bardiya district as a whole, undertaken in chapter 4, also shows that the policy process was constantly reflected on by district-level political leaders who felt ambivalent about equating "participation" with the local *begaari* tradition. The vulnerability of the corvée standard at the district level, combined with the village-level remolding of the *begaari* modality, provided fertile soils for marginal villagers to engage in "cultural politics" over the way that "participatory" river control was put into action.

Just as was concluded at the end of chapter 5, the decline of the "participatory" river control policy in Bardiya towards the end of the 1990s can be attributed to the constant renegotiations of the overall policy direction at the district level, and also to the contested nature of the grassroots policy implementation. It is true it would be possible to make sense of the downward trend of the policy in the light of the changed political contexts at the national and district levels. In the eyes of central political leaders, the value of the policy had decreased as an instrument for patronage distribution, owing to the disparity that had emerged between the party in power at the center and that dominant at the district level. However, this type of "descending" analysis, more commonly used among policy analysts than is the "ascending" approach, does not highlight the entirety of the dynamics surrounding the unfolding of the "participatory" river control policy. Indeed, as discussed in chapter 3, the nepotistic use of the policy was far from being a foregone conclusion at the center, and was constantly reflected on among central policymakers, who felt ambivalent about reducing river control to the delivery of pre-packaged projects. The following concluding chapter brings together the results of the national-level and district-wise analyses of the policy process, and also of the two village-level case studies, with a view to gaining further insights into the policy process.

# Notes

1. As explained in chapter 3, the VDC was divided into nine electoral districts, in each of which a Ward Committee was formed. The Ward Committee was composed of one chair and four members. One of the four ward memberships was reserved for a female representative. The ward chair also served as a member of the VDC. The VDC chair, the vice-chair, and all the ward representatives were directly elected by residents.

2. There existed three households of mixed marriage, two between *Tharu* males and *Pahadi* females, and the other between a *Dheshi* male and a *Tharu* woman. All these couples lived in *Tharu*-only settlements, and are counted as *Tharu* households.

3. *Dheshi* is an abbreviation of *Madheshi*, which literally means people living in the southern plain. According to the original meaning of the word, *Tharus* should therefore be a sub-group of *Dheshis*. However, the two were distinguished in the case study village.

4. The Nepali Congress (NC) and the *Rashtriya Prajantra* Party (RPP) were more or less of equal size in their support bases. Judging from the poll of the local elections held in 1992 and 1997, supporters of the CPN-UML did not constitute an absolute majority. This posed a difficulty for the CPN-UML in executing their decisions in the face of obstructions by the greater part of the population in support of the NC and the RPP. The Nepal *Sadbhavana* Party (NSP) also commanded some albeit the smallest influence in the village.

5. Moreover, *Tharus* were increasingly given opportunities to head user committees for public works projects in the 1990s. The elected leaders from the CPN-UML had to consider the widespread grievances against *begaari* that *Tharus*, especially those from opposition parties, were able to build on, to frustrate the project implementation.

6. For example, with the support of the CPN-UML, a group of *Tharu sukumbasis* obtained permission from the government to stay on the land. Similarly, residents of two *Pahadi sukumbasi* areas, where their landed neighbors had formed their forest group in the early 1990s to the exclusion of the *sukumbasis*, subsequently managed to secure their own forests with the help of local politicians.

7. For example, when one *dalit* settlement was told to build a trail passing through their settlement, its residents generally resented having to solely bear the burden for the path construction that would also benefit other landed villagers. At the same time, they took it as an opportunity to counter their neighbors' intolerance, and subsequently formed a user committee to complete the trail by contributing their labor.

8. For example, in one settlement where *Pahadis* and *Tharus* coexisted side by side, *Tharus* managed to get *Pahadis* to agree to work side by side with *Tharus*, albeit only for one day, in their annual five-day trail maintenance activities. However, given a low turnout by *Pahadis*, *Tharus* finally refused to clean the road. In 1999, two *Tharu* settlements refused to continue providing unpaid labor during the construction of a school. They were no longer able to bear a low turnout from *Pahadi* communities. *Pahadis* from one particular

settlement had not even turned up at all. Two other *Tharu* settlements subsequently followed suit.

9. From 1998 to 2002, a faction within the CPN-UML formed its breakaway party CPN-ML (-Marxist and Leninist) in Nepal. In Pyauli, all the four ward members who won the 1997 local election were from the CPN-UML, two of whom subsequently defected to the CPN-ML in 1998. The ward chair continued to be affiliated with the CPN-UML.

10. Most of them went around their settlements to request the residents to provide labor, but rarely made extra endeavors, for example, to get their neighbors to assemble in one place with a view to leading them to the construction site. Others even neglected to mobilize their settlements, and in such cases, the ward chair himself had to ask those defaulters to get them to make rounds in their localities.

11. *Tharu* and *Pahadis* lived in distinctly separate settlements, except for two localities where both the groups resided side by side. In the two places, the user committee made a request through the *Tharu badghars* or leaders, who appealed to both the ethnic groups to participate in the project. However, practically all *Pahadis* in these two areas ignored the *badghars*.

12. Excerpt from my interview of March 2001.

13. See Footnote 7.

14. Excerpts from my interview of February 2001.

15. As stated in chapter 1, in Pyauli, a sample of about 150 households was chosen from the voter lists. For this, I first grouped voters by settlements, and used a random number table to select a sample from each settlement.

16. Excerpts from my interview of April 2001.

17. Of all the (forty-eight) sample households, nearly thirty percent (nineteen households) did not send any family members to the project site on behalf of their forest groups. This can be attributed to discords and disputes arising out of party rivalries, or out of various social and economic differences, such as those described in the main text.

18. For example, in the second case, the *Tharus* and the *Pahadis* were at odds with each other because the two groups constituted support bases of different political parties. In the case of the first example, leaders from the NC who were landed *Pahadis* were also inclined to take advantage of the project to display their control over their members who were *sukumbasis*.

19. While many of the chairs of other wards did not heed the request, even those ward chairs that mobilized labor from their respective constituencies brought fewer than twenty persons, who did not spend more than a few hours there. A number of government officials spent their time resting in the shade, and some even sneaked away from the site in broad daylight. Many school children frolicked in the water. Moreover, these groups from outside Pyauli tended to focus on filling sand in sacks, thereby leaving aside the more physically demanding work of carrying the sandbags to the riverside, and piling them up in the river.

20. It is to be noted that the local community was far from being homogenous or cohesive, but was inhabited by *Tharus* of different social and economic status. There was no uniformity in the manner that different groups participated in the project. Some of the landless peasants were compelled to participate virtually on a daily basis by their landlords whose land was at risk. Other residents, on the other hand, generally participated on a rotational basis. At the same time, even those landowners, who offered tenancies to other peasants, provided labor at the construction site. This was partly because their land was in

imminent peril, but also because they had been used to being involved in *begaari*, unlike their *Pahadi* counterparts. In this exclusively *Tharu* area, as in the other *Tharu* localities in Pyauli, landlords, and their tenants or laborers alike were engaged in *begaari* tasks, to a greater or lesser degree.

21. Excerpts from my interview of September 2000.

22. Michel Foucault, "Afterword: The Subject and Power," in *Michel Foucault: Beyond Structuralism and Hermeneutics*, ed. Hubert L. Dreyfus and Paul Rabinow (Brighton: The Harvester Press, 1982), 211.

23. Michel Foucault, "Two Lectures," in *Power/Knowledge: Selected Interviews and Other Writings 1972-1977 Michel Foucault*, ed. Colin Gordon (New York: Harvester Wheatsheaf, 1980), 98-99.

# Chapter 7
# Conclusions and Implications

This study set out to analyze the policy process of "participatory" river control in Nepal by tracing its emergence and evolution in the capital, as well as its local manifestations at the district and village levels. As described in the introductory chapter, its main aim was to gain a nuanced understanding of the power dynamics surrounding popular participation in policymaking, by overcoming the weaknesses of existing studies that tended to assign fixed subject positions to high-level policy "elites", and to overlook the leverage their "clients" can potentially exercise over the unfolding of the policy process. In this regard, the study did not start with abstract reifications of actors, but instead made empirical inquiries into particular practices, and addressed three broad questions. First, the research inquired into how the "participatory" river control policy came into being at the national level. Second, it examined how the top-down directive of "participatory" river control intermeshed with the life worlds of local actors in the Bardiya district. Finally, the study analyzed the unfolding of the "participatory" river control policy, as a meshing of "cultural politics" that various stakeholders engaged in, ranging from villagers to central politicians. This concluding chapter first summarizes the major findings from the different empirical chapters. It then teases out the theoretical implications of the findings, by referring to existing studies advocating public engagement with policymaking. This chapter concludes by drawing lessons for practical politics promoting people's engagement with the policy process.

## Empirical Findings

This section discusses the empirical findings of this study, paying special attention to power struggles and the policy process. It first focuses on power relations in the case study villages, in view of the hypothesis of this study that local actors were not simply subsumed under the "official" policy discourse, but exercised some leverage over the policy process. In line with this hypothesis, village politics transcended immediate localities and impacted on the wider policy process. This can also be attributed, as described in the second part of this section, to the negotiated nature of the policy process at the national and the district level. The manner in which "participatory" river control was translated into action was ceaselessly reflected on among policymakers who held fluid and contingent subject positions. Based on the findings concerning village power dynamics and policymaking at higher levels, this section concludes by exploring alternative ways of explaining popular participation in the policy process. The ceaseless renegotiations over the policy direction at the national and district levels, combined with the contested nature of village politics, provided a fertile ground for the general public to exercise influence over the unfolding of "participatory" river control.

## Oppression/Resistance Nexus in Local Power Dynamics

As explained in chapter 2, this study has drawn on the two strands of debates on power, namely the "structuration" perspective, and the Foucauldian notion, both of which shed light on the "entanglements" of oppression and resistance.[1] According to the conventional "power as domination" notion, power is attributed to influential actors, who exercise it over others in such a manner as to harm others' interests. As described in chapters 5 and 6, this mainstream view, which pits resistance by "oppressed" groups against control by "dominant" players, does not elucidate village situations in which the two were interwoven with each other in daily social interactions, and in which ongoing social relations were continually being readjusted. To capture the subtle and intricate village power contestations, it is crucial to draw on the "structuration" perspective and the Foucauldian conception, as explained below.

*"Structuration" View of Power*

As stated in chapters 5 and 6, village power struggles cannot be explained solely on the basis of the "power as domination" view, but can be grasped within the "structuration" view that regards power to be immanent in the daily flow of social interactions. In both of the case study areas, the hierarchical social orders not only constrained the thoughts and actions of oppressed groups, but also provided grounds for them to capitalize on the inequalities embedded in society to renegotiate ongoing power relations. The oppressed in the two localities were constantly seeking opportunities to turn "structural" constraints into "structural properties" in response to both the limitations and opportunities arising in their day-to-day lives.

In order for such "duality of structure" to take effect, "infra-politics"[2] played an important role, whereby marginal actors engaged in undeclared, disguised resistance in an attempt to unobtrusively re-negotiate power relations. In both of the villages, *Tharus*, *sukumbasis*, *dalits*, and women tried out various ways of tacitly and subtly easing up on their contributions. Day-to-day social interactions, through which inequitable social relations had been ceaselessly renegotiated, had accorded the underprivileged what Bourdieu terms "habitus",[3] that is they had gained a practical understanding of the feasibility of slackening off their work covertly, and at the same time, of the difficulty of uprooting deeply entrenched social practices.

What facilitated such covert resistance was the increasing complexity of politics that had come to be anchored around party, ethnic, class, gender, and other social and economic differences. As various aspects of village struggles came to be interwoven to bring about increasingly dynamic and complex social relations, the subject positions of social actors became less neatly defined and more fluid, multifaceted and fragmented. The growing intricacy of village politics gave rise to opportunities for the oppressed to step up their resistance to existing social relations. In this respect, the party politics, that came into the open in the 1990s, often played a crucial part by refracting with other social tensions in such a manner as to encourage the oppressed to challenge social inequity, with recourse to political parties competing with one another for votes.

## Foucauldian Notion of Power

The "duality of structure", described above, has important implications for how the Foucauldian type of power operated in the case study localities. In Bardiya, public works projects were historically undertaken drawing on the corvée tradition called *begaari*. The inequality embedded in the corvée practice was justified with reference to a version of local history that depicted migrant *Pahadis* as agents that had brought development into the backwater originally inhabited by *Tharus*. Moreover, the decentralization and participation (D&P) rhetoric was often used by politicians as a euphemism for *begaari*. The corvée standard emanated what Foucault terms "disciplinary" power to constrict the perceptions of villagers in a manner to incline them to take for granted the application of *begaari* for public works. According to Foucault, at the same time, different social actors attribute different meanings to social norms, and constantly renegotiate and modify "disciplinary" power. It is therefore imperative to pay attention to "antagonisms of strategies",[4] that is how different actors in particular situations are subjugated to and resist "disciplinary" power, as attested to by the two case studies.

In both of the case study villages, the *begaari* standard in itself did not succeed in bringing individuals to provide unpaid labor submissively, and had to be ceaselessly reworked to maintain its hold over the areas. This arose from the "open texturedness" of discursive practices, that is the inability of discourse to impose order with finality. The self-claimed righteousness as the "change agent", referred to above, allowed *Pahadis* to exact the free labor of *Tharus*, and at the same time, compelled them to show leniency, as "benevolent patrons", towards the plight of the latter group. The types of reasoning of *Pahadi* dominance, were thereby tantamount to a double-edged argument, serving both to uphold and to negate it. The "disciplinary" norm of the corvée practice, articulated through such contingent argumentation, was inherently vulnerable to opposing assertions that point to the need to relax *begaari* obligations.

What added to the absence of the tenability of the *begaari* standard was the "dual" nature of ongoing social relations, explained in the preceding section. As the "duality of structure" intensified village power struggles, the taken-for--granted corvée norm that had long been central to village politics was also shaken. The decline of the *begaari* norm in turn affected interpersonal relations, in that it inhibited *Pahadis* from unreservedly pushing *begaari* ahead, while allowing *Tharus* to take advantage of the "duality of structure." In this way, the precariousness of the "disciplinary" corvée norm, highlighted by the Foucauldian conception, and the "duality" of the inequitable social orders, pointed out by Giddens' "structuration" view, provided fertile soils for each other, in rendering daily power contestation a dynamic and intricate process. Before analyzing the implications of such grassroots power dynamics on the policy process, the following section presents the findings about policymakers.

## Shifting Subject Positions of Policymakers

As described in chapters 3 and 4, the discursive practices of policymakers, at the national and district levels, were far from being monolithic and immutable, but were fluid and incoherent. Given their overall propensities to equate development with a centrally directed process of infusing exogenous knowledge and resources into outlying areas, it can be surmised that national policymakers merely relegated "participatory" river control to the delivery of pre-packaged projects. Moreover, because "participatory" river control was translated into action largely as corvée-based endeavors, politicians in the Bardiya district seemingly sought to appropriate the local policy process to blatantly uphold the local corvée practice. These ostensibly unquestionable accounts, however, fail to capture the "open texturedness" of policymakers' discursive practices. Moreover, owing to the precariousness of their articulatory practices, policymakers did not maintain a stable and coherent identity but faced "social antagonism."

*"Fragmentation" in Discursive Field*

In Nepal, the D&P policy, which came to the fore in the 1990s, was largely relegated to a centrally directed process of infusing external knowledge and re-sources into villages. As stated in chapter 3, the D&P rhetoric spilt over into the "participatory" river control policy, in such a way as to equate flood mitigation with the delivery of handouts to the local populace, while giving an illusion of overhauling the top-down conduct of development. Accordingly, politicians and bureaucrats tended to equate river control with a centrally directed process of distributing engineering solutions, while reducing participation to an instrument to ensure people's acceptance of their projects, and to solicit their labor contributions.

This ostensibly reasoned explanation is flawed in that it is founded on a "repressive hypothesis" that policymakers only seek to entrench their clients' dependence. Such a proposition draws on a binary logic that simplistically categorizes social actors into "goodies" and "baddies", and thus provides only a one--sided view of the policy process. As explained in chapters 3 and 4, the constrictive notion of "participatory" river control was ceaselessly reflected on among policymakers, both at the national and district levels, whose discursive fields were not monolithic, but contained heterogeneous and contradictory elements. For example, there existed political leaders who pointed to the need to address broader socio-political issues, such as India's undue interventions in Nepal's river control, and social exclusion that forced marginal people to reside in flood-risk areas. Similarly, some bureaucrats proposed to do away with perfunctory public involvement in "participatory" river control, and to enhance the capacity of community groups to formulate their own projects.

Such "fragmentation" within the discursive fields of policymakers emanated from the inherent impossibility of forcing a closure to their discursive practices. Their mainstream view of "participatory" river control had to distance itself, in order to attain its identity, from other plausible stories that point to the precariousness of the widely held notion. The "orthodox" account was thereby vulnerable to other competing discourses. For example, local politicians in Bardiya generally equated "participatory" river control with the corvée-based construction of physical structures, on the grounds that *Pahadi* migrants had historically guided uneducated *Tharus* to development, by exacting labor contributions from the latter. This reasoning projected *Pahadis* as "benevolent benefactors", and thus was inherently vulnerable to counter-arguments that point to the downsides of the corvée practice. They included an argumentation that small-scale structures built by *Tharus* could not withstand the forces of severe flooding, and an assertion that denounced the corvée tradition for forcing *Tharus* to forgo their own work for days in spite of their hand-to-mouth way of life.

### "Undecidability" of Social Identity

Since these types of counter-arguments were equally plausible, politicians in Bardiya usually felt ambivalent about whether "participatory" river control should be reduced to corvée-based endeavors. This is what Laclau and Mouffe call "social antagonism",[5] which denotes the impossibility for social actors to attain fully coherent identities in the face of the "undecidability" of their articulatory practices. Accordingly, chapters 3 and 4 not only demonstrate that, both at the national and district levels, different policymakers maintained divergent discursive practices, but also that there existed rifts in the subjectivity of each policymaker, who normally held multiple and fragmented subject positions. For example, as described in chapter 3, there existed central politicians whose discursive practices embodied incoherent elements, such as those relegating "participatory" river control to a pre-packaged project, while at the same time, stressing the need to grant greater local autonomy.

This does not mean, however, that the articulatory practices of policymakers were wholly "undecidable" without any enduring process. On the contrary, because of the contingency of discursive practices, there needed to be a platform that allowed various policymakers to dissolve their differing and unstable perceptions and to communicate with one another. Policymakers thus tended to coalesce around the above "mainstream" view equating "participatory" river control with the distribution of pre-packaged projects. The discursive field of policymakers exhibited a certain degree of "regularity in dispersion" in that the "orthodox" account of river control permeated their articulatory practices irrespective of their subject positions.

At the same time, as repeatedly stressed in this and the preceding sections, individual officials did not have fixed identities, as opposed to conventional political analysis in which individuals are presumed to have given interests. An important implication of the fluid and inconsistent identities of policymakers was that one should avoid prejudging the likely course of the unfolding of a centrally directed policy, however immutable and imposing it may seem. The policymaking of river control was ostensibly the unilateral process of relegating "participatory" flood mitigation to the delivery of predetermined "technical" packages. At the same time, when the actual articulatory practices of policymakers are placed under scrutiny, as is done in chapters 3 and 4, it is evident that the policy process was subject to ceaseless reflections among policymakers with multiple and incoherent subject positions. The following section considers how to make sense of popular participation in the policy process, in view of the contested nature of the unfolding of the "participatory" river control policy.

## Popular Participation Immanent in Policymaking

The policy process of "participatory" river control ostensibly worked to inhibit the participation of the flood-affected populace, in that it was largely relegated to the delivery of the pre-determined package. At the same time, as stated thus far in this chapter, "target groups" were not necessarily submissive to the manipulative process of equating "participatory" river control with corvée-based undertakings, while policymakers did not always seek to defend their influence in the delivery of river control resources in rural areas. The contingent nature of both the village-level power dynamics, as well as the central and local-level policy processes, indicates that there existed a fertile ground for people's engagement in the unfolding of the "participatory" river control policy. To ascertain the immanency of popular participation in the daily flow of policymaking, it is crucial to adopt an "ascending" analysis that examines the "cultural politics" that the spectrum of stakeholders engage in, ranging from target groups to policy elites, to ceaselessly reflect on and renegotiate the unfolding of policymaking.

*"Ascending" Analysis*

The "participatory" river control policy came to be ignored by the government towards the end of the 1990s. In the latter half of the 1990s, the central government increasingly gave instructions to use contractors in lieu of labor contributions, and specified the locations and types of river control structures, instead of sending a blanket authorization for each district to use the budget at will. This can be attributed to the decline in the value of the policy as a means of patronage distribution, from the viewpoint of national politicians, given the emergent discrepancies between the political party in power at the center, and that dominant at the district level. Such an account of the downward trend of the "participatory" river control policy is tantamount to what Foucault terms "descending" analysis,[6] which examines who possesses power, and what strategies they employ to pursue their own agenda. In the case of this study, analysts with a "descending" view would seek to identify who were in official positions of framing "participatory" river control, and to look into how those policymakers sought to use the river control resource to strengthen their influence in the policy process.

As already stated in this chapter, however, this book shows that policymaking was played out through informal micro-power struggles among the entire spectrum of stakeholders including policy "clients", not solely through institutional power exercised by a limited number of those in privileged positions. This resonates with a call by Foucault not only to examine power from the angle of institutions, but also to adopt an "ascending" strategy to focus on the inconspicuous micro-power experienced by different players.[7] The "ascending" analysis of the two case study villages, undertaken in chapters 5 and 6, illustrates that, even under "dictatorial" circumstances, the policy process continued to be reshaped and remade by ongoing power dynamics, through which "marginal" segments of the flood-affected populace resisted and struggled to rework the top-down directive. As a result, as described in the two chapters, it was infeasible to put corvée into action without being entangled in ceaseless power struggles among various local actors over the inequitable tradition. In this way, policy "clients" in Bardiya were eventually able to exercise leverage over the decline of the "participatory" river control policy towards the end of the 1990s.

At the same time, in making sense of the overarching policy process, it does not suffice to move upward geographically, namely to examine the micro-power struggles by grassroots policy "clients." An "ascending" analysis should also focus on the politicians and bureaucrats, both at central and district levels, who held multiple and incoherent subject positions. The fragmented nature of their subjectivities also laid the ground for villagers to exercise influence over the policy process. As stated already in this chapter, despite their overarching propensities to relegate "participatory" river control to the delivery of pre-packaged technical solutions, policymakers' stances ceaselessly wavered between their propensity to seek to tighten their grip on river control resources, and their inclinations to cater to the public. It is therefore crucial to refrain from regarding policymakers as a monolithic entity seeking to maintain their hold on the policymaking process. The downward trend of "participatory" river control in the 1990s arose not only as a result of the popular resistance, but also because of the transient and inconsistent nature of policy elites that served to open up space for the policy change.

*"Cultural Politics"*

The relevance of the "ascending" approach to policy analysis, described above, corroborates the notion of "cultural politics" propagated by the "anthropology of policy", which, in the field of development studies, is adopted by "post-development" thinkers. According to this strand of policy studies, as explained in chapter 2, policymaking is implicated in the complex meshing of varying meanings attached by different stakeholders, whose subject positions are fluid and inconsistent. This does not mean that the "anthropology of policy" downplays the oppressive nature of the policymaking process. On the contrary, the "anthropology of policy" regards policy narratives as constitutive of what count as "knowledge" in such a manner as to "discipline" the general public to embrace the expansion of state intervention, or the "governmentalization" of society. At the same time, as pointed out by Foucault, such "disciplinary" power is not entirely oppressive, but is also productive in that it facilitates policy stakeholders to respond to and renegotiate the norm of "governmentality." To counteract the manipulative nature of policy interventions, therefore, proponents of the "anthropology of policy" turn to the "utopian possibility of reconceiving and reconstructing the world from the perspective of, and along with, those subaltern groups that continue to enact a *cultural politics* of difference."[8]

In line with this assertion of the "anthropology of policy", this study shows that popular engagement with policymaking was immanent in the daily flow of the policy process of "participatory" river control, although it was relegated largely to the delivery of pre-packaged projects. The policy process did not entirely inhibit the participation of the flood-affected populace, but also prompted them to engage in "cultural politics" to renegotiate the definition of "participatory" river control. National and district-level policymakers, too, engaged in "cultural politics" to ceaselessly reflect on how the policy was executed. The "cultural politics" among the range of stakeholders rendered policymaking, in however top-down a manner it might have been executed, a highly contested process that opened up space for policy "clients" to exercise influence over the unfolding of the "participatory" river control, consequently leading to the decline of the policy in the latter half of the 1990s. The following section considers the theoretical implication of such "cultural politics" for existing studies on popular participation, which tend to assume the necessity of deliberate external interventions to activate public participation in policymaking.

## Theoretical Implications

The preceding sections summarized the empirical findings of this study, with a focus upon the nature of power dynamics and the policy process. First, local power dynamics were not only oppressive but were also productive in that inequitable social orders both constrained and facilitated resistance by marginal actors. Second, the manner in which the "participatory" river control policy was executed was ceaselessly reflected on among policymakers who held fragmented and incoherent subject positions. Finally, an "ascending" perspective, examining the power struggles experienced by both high-level and grassroots stakeholders, points to the existence of a fertile ground for policy "clients" to exert leverage over the unfolding of the "participatory" river control policy, and the ultimate decline of the policy towards the end of the 1990s.

This section considers theoretical implications of these empirical findings, in relation to research on popular participation in the policy process. For this purpose, it focuses on development studies, and draws lessons for those studies that suggest moving away from the orthodox "user group" approach towards "citizen" participation, that is broader people's engagement with the policy decisions affecting their lives. The advocacy of external interventions to activate "citizen" participation is scrutinized, in view of the emancipative potential immanent in the daily flow of policymaking.

### Rethinking "Citizen-Subject" Participation

As stated in chapters 1 and 2, there are increasing calls among researchers on participatory development, to move away from the conventional "user group" focus, in favor of the promotion of "citizenship" for broader popular participation in the policy process. "Citizen" participation has emerged as a converging concern among those researchers who advocate public engagement with policymaking. One weakness of this strand of research is that it disregards the potentiality inherent in the policy process to accord people some leverage over its unfolding. It acknowledges the popular agency of "target groups" to shape how a policy evolves on the ground, but not how it is made at higher levels. In their views, it is therefore necessary to launch a distinct set of initiatives to turn the general public into fully-fledged "citizens" capable of engaging in policy dialogues. "Citizen" participation, catalyzing changes in broader decision-making processes, according to its protagonists, is diametrically opposed to "user group" participation that only allows people to take on externally supported projects.

However, the empirical examinations made in this study indicate that, contrary to the view held by the proponents of "citizen" participation, the general public are not necessarily "barred" from policymaking, but are already "citizens" potentially capable of influencing the process. A main implication of this for research on people's participation is that one should avoid focusing solely on the impacts of external programs in immediate localities, and should also heed their long-term, far-flung repercussions, a point which is also raised by Williams.[9] For this purpose, protagonists of "citizen" participation should assimilate themselves to the "anthropology of policy" that locates popular agency in a complex web of "cultural politics" taking place in myriad places. This would allow proponents of "citizen" participation to overcome the above-mentioned weakness, namely their tendency to play down beneficiaries' capacity to exert leverage over the overarching process of policymaking. Participatory intervention for a local community does not take place in a bounded, clearly demarcated space, the impacts of which spread to a multitude of interrelating and overlapping sites.

Another main implication for the research on participatory development is that one cannot necessarily regard "citizen" participation as the antithesis of "user group" participation. According to its proponents, "citizen" participation marks a radical departure from the conventional "user group" approach, in the sense that people are not relegated to "users and choosers" of external programs, but are enabled to act as "makers and shapers" who catalyze changes in broader sociopolitical circumstances, to use the phrases coined by Cornwall and Gaventa.[10] Underlying this seemingly renewed notion of "citizen" participation is the assumption that people would not be "citizens" under the "user group" approach. However, this presumption does not hold, as attested to by the case of the "participatory" river control policy, under which people were reduced to "consumers" of pre-packaged projects, but were able to influence higher-level decision-making, The advocacy of "citizen" engagement with the policy process, however well-intentioned it may be, is therefore liable to inhibit people's autonomy and spontaneity in shaping their strategies, and instead to orient them to accept that they would not be "citizens" without outside assistance. Its beneficiaries would thus remain "users and choosers" of external programs, just as under the conventional "user-group" approach.

A "citizen", as is conceived by protagonists of "citizen" participation, is tantamount to a "subject" who is subjugated to oppressive power, whom Cruikshank succinctly calls a "citizen-subject" as opposed to a "citizen/subject."[11] To break with the orthodox approach to public participation, one should take care not to relegate external interventions to "technologies of citizenship"[12] that facilitate outside agencies to pursue their agenda in lieu of the ones shaped by beneficiaries. It is imperative, instead, to acknowledge that the potentiality of active public engagement in policy debates is inherent in the ongoing flow of the policy process, and thereby to pay attention to the popular agency immanent in existing societal forces, as distinct from participation in imminent interventions. The following section draws out some lessons for those seeking to translate into action such more spontaneous, less formulaic approaches to promoting greater public participation in policymaking.

# Practical Considerations

How should external agencies assist in the promotion of popular participation in policymaking, while avoiding constituting and regulating the subjectivities of their "target groups"? As explained in the preceding section, proponents of the ostensibly renewed strategy of activating "citizen" participation still tend to constrain people to act as "users and choosers" of externally conceived programs. An alternative approach to "citizen" participation is therefore called for, which allows the general public to become "makers and shapers" of their projects to influence the policy decisions affecting their lives. For this purpose, this section examines an anecdote from Majuwa that illustrates the relevance of this study to practical politics promoting people's engagement with the policy process.

## Lessons from an Anecdote in Majuwa

In Majuwa, one donor agency launched another river control project in January 2001. The project originally envisaged the mobilization of unpaid labor, drawing on the corvée tradition. At the time, however, another road construction project was about to be launched with villagers' *begaari* contributions. Moreover, at the onset of the monsoon, the busy season for farmers was fast approaching. This provided *Tharu* peasants with grounds to openly refuse to offer *begaari*, thereby preventing the project from being started. As the donor agency and the elected representatives of the village, namely the VDC chair and the ward chair, came under increasing pressure to initiate the construction of flood control structures, it was decided to assign the corvée obligation to landowners instead of passing it to their tenants. Landlords, a vast majority of whom were *Pahadis*, had to pay a market-rate wage to their *Tharu* tenants, if they were to depute the latter to provide labor. This amendment struck me as a major departure from the past, when corvée had been the responsibility of tenants, not their landlords.

Through my fieldwork for this study, I had built up rapport with a number of *Tharu* peasants, who were pleased to divulge stories of the historical plight of *Tharus* and their daily struggles. I was expecting that those informants would happily describe the news as a major breakthrough in their resistance to the ongoing social order. However, none of them regarded this revision as a special step forward, but saw it as another incident in their continual struggle to redress social

injustice. In the past, as demonstrated in chapter 5, the mode of the corvée practice had ceaselessly been renegotiated and modified, given the intermingling of
oppression and transformation in daily social interactions that accorded the general public the latitude to influence the way that the corvée tradition was put into
action. The alteration of the donor project plan evolved out of a series of those
micro-power struggles that had subjected the corvée tradition to continual readjustments. Through years of assiduous resistance, *Tharu* peasants managed to
obtain remuneration, albeit at rare intervals, as exemplified by the experiences of
the river control projects in the mid-1990s. In this sense, the "new rule" of corvée
was just an extension of the few exceptional cases that had occasionally occurred
despite the historically predominant pattern of non-payment for *Tharus'* labor
contributions.[13]

A question that follows from the theoretical implications, provided in the
preceding section, is what role an external agent can play to catalyze policy
changes in donor programs and projects to enable ordinary people to take greater
control over them. A general policy of the donor was to create community groups
to plan and manage projects, and this was also applied to the 2001 undertaking. In
line with the general trend among donors, this agency also tended to reduce
"participation" to the procedural matter of prompting beneficiaries to organize
themselves and to translate into action a sequence of predetermined activities,
namely the orthodox "user group" approach of getting people to take on the
execution of externally conceived programs, rather than promoting broader
popular involvement in deciding the overall direction of the project. A proponent
of "citizen" participation would advocate starting a separate process of countermeasures against the donor policy, with a view to turning *Tharus* into fully-
fledged "citizens" engaging in policy dialogues with the donor.

The case of the 2001 project attests to the contrary. The adjustment to the
donor policy evolved out of the daily struggles of *Tharu* peasants to renegotiate
and challenge the corvée practice. Even under the "user group" approach, *Tharus*
managed to persuade the donor to allow villagers to redraw the overall plan of the
project in defiance of the donor policy to focus on getting people to follow
pre-packaged "participatory" procedures. Contrary to the presumption widely
held among protagonists of "citizen" participation, it was not a prerequisite for
the disadvantaged to enter into dialogues with donor representatives, in order for
them to exercise leverage over the unfolding of the donor policy. This anecdote
thus illustrates that a potential for the emancipation of marginal people, who are
seemingly denied participation, is inherent in the daily flow of the policy process.

This does not mean that there was no scope for advocacy against the do-
nor-wide tendency to follow a preset sequence of activities. At the same time, it
would have been a grave mistake to launch a campaign against the donor policy,
while blocking the implementation of the project. In such a scenario, villagers and
donor personnel would become contrasted as "goodies" and "baddies", thus un-
necessarily alienating the latter from the potential of making self-imposed ad-

justments to the "participatory" rule. In reality, the dispute that arose during the 2001 project did compel donor personnel, who also held fluid and multiple subject positions on his organization's policy,[14] to tend to leniency towards the plight of *Tharus*. In the context of this project, therefore, it was imperative to start any campaign against the donor policy, only after ascertaining how the donor and villagers worked out their differences regarding the project. It was crucial to pay due regard to, and to capitalize on the potentiality of the "underprivileged" to participate in shaping the unfolding of the policy process even though they were ostensibly marginalized in decision-making processes.

## Building on "Cultural Politics" Inherent in Policymaking

In concluding this study, I put together its implications for proponents of greater public engagement in policy decisions. The case of "participatory" river control in Nepal indicates that there exists a fertile ground for policy "clients" to influence the policy process, which takes place in a complex web of "cultural politics" occurring in myriad places. In supporting public engagement with policymaking, therefore, external agents should not solely be concerned with deliberately bringing citizens' voices to the attention of policymakers. It is imperative to start out by considering how the ongoing flow of social interactions, in which domination and emancipation are intermingled, can potentially play a part in opening up space for ordinary people to participate in the policy process. By assessing the opportunities and limitations arising from interactions among various policy stakeholders, it is feasible to devise strategies that both build on, and complement struggles by marginal groups to engage in the decision-making affecting their livelihoods. Similarly, as summarized earlier in this chapter, this study shows that policy elites do not necessarily seek to defend and expand their influence in the policy process, but that their subjectivities are transient and multifaceted. In order to take advantage of the fluid and inconsistent nature of officials' subject positions, those involved in practical politics should avoid assuming in advance that policymakers seek to suppress the general public, thus preempting the space for the two to find a common ground.

Past guidelines on "participatory" interventions were overly concerned with the formal process of designing, executing, and evaluating action programs. This is also the case with the Participatory Poverty Assessments (PPA), which came to the fore around the turn of the millennium as a strategy to facilitate the participation of the poor in policymaking. According to one book published by its main proponent, the World Bank, "(i)nstead of a predetermined set of questions . . . PPAs use a variety of flexible methods that combine both visual . . . and verbal . . . techniques."[15] While it is true that its practitioners choose among a wide range of tools, the PPA proceeds in a step-by-step sequence of agenda-setting through field research, broad-based consultations with a cross-section of civil society, and decision-making by policymakers. In this sense, the PPA is founded on the orderly world-views, referred to at the start of this book, that high-level decision-makers are liable to impose policies incongruent with local circumstances, and that it is crucial to launch a separate set of countermeasures against elite-controlled processes of policymaking.

Such a linear, purposive, "logical" view of the policy process, however, does not tally with the contingent and unpredictable nature of policymaking that is implicated in dynamic interplays of "cultural politics" among different stake-holders. It is imperative for those involved in policy advocacy, to adopt less structured, more informal investigations, to gain a nuanced understanding of the ongoing renegotiations of power, and then to design strategies accordingly. As pointed out by Cleaver, one should avoid assuming that "the practice of partici-pation is unilinear and cumulative."[16] As long as external agents engage in a set of goal-oriented activities with the predetermined yardstick to bring people's per-ceptions to the attention of policymakers, they unwittingly seek to get their beneficiaries to follow externally conceived agendas. By assuming that the general public are subjugated to the oppressive and manipulative nature of policymaking, outside agencies inhibit the self-reflexive nature of the daily flow of policymaking, in which both policy "elites" and their "clients" ceaselessly re-flect on the manner in which policies are executed. The importance of avoiding the projection of one's unsubstantiated presumptions onto the subtle and intricate social interactions among various stakeholders is also exemplified by the case study of the donor-supported project, taken up in the preceding section.

Bourdieu cautions that social scientists risk projecting logic onto non-theo-retical flows of social interactions.[17] Bourdieu picks up a calendar as a metaphor of scientific construction. A calendar "substitutes a linear, homogenous, continuous time for practical time which is made up of islands of incommen-surable duration, each with its own rhythm, a time that races or drags, depending on what one is doing."[18] One should bear in mind that, by virtue of attributing formal properties to the uncertainty surrounding the policy process, "scientific" analyses are liable to divert attention away from the transient nature of the actual practices of policymaking.

To exercise reflexive vigilance against the potential risk of imposing regu-larities onto informal flows of social interactions, it is imperative to undertake ethnographic investigations into the "cultural politics" among different stake-holders, which renders policymaking a contingent and unpredictable process. Given their commitment to take the side of the disadvantaged, proponents of popular participation succumb to the "repressive hypotheses" that policymaking is preempted by policy "elites", and that policy "clients" exercise little leverage over its unfolding. Instead, it is imperative for "well-meaning" outsiders to conduct the "ascending" mode of ethnography, which delves into the intermin-gling of oppression and emancipation in daily social interactions, the concomi-tantly fluid and incoherent nature of one's subject positions, and the resultant impossibility for policymaking to unfold in such a predictable manner, as can be deduced from the structural locations of various stakeholders. In order to avoid imposing unfounded presumptions, external agents should engage in the "reflexive" methodology, referred to in the introductory chapter, while conceiving how they should assist ordinary people in their engagement with the policy process. It will not be possible for outside agencies to grasp the contingent and spontaneous nature of policymaking, without constantly reflecting on what they consider would activate popular participation in the policy process.

## Notes

1. Joanne P. Sharp, Paul Routledge, Chris Philo and Ronan Paddison, "Entanglements of Power: Geographies of Domination/Resistance," in *Entanglements of Power: Geographies of Domination/Resistance*, eds. Joanne P. Sharp, Routledge Paul, Chris Philo and Ronan Paddison (London: Routledge, 2000), 1-42.

2. James C. Scott, *Domination and the Arts of Resistance: Hidden Transcripts* (New Haven: Yale University Press, 1990).

3. Pierre Bourdieu, *Outline of a Theory of Practice* (Cambridge: Cambridge University Press, 1977).

4. Michel Foucault, "Afterword: The Subject and Power," in *Michel Foucault: Beyond Structuralism and Hermeneutics*, ed. Hubert L. Dreyfus and Paul Rabinow (Brighton: The Harvester Press, 1982), 211.

5. Ernesto Laclau and Chantal Mouffe, *Hegemony and Socialist Strategy*. 2d ed. (London: Verso, 2001), 122.

6. Foucault, "Afterword," 222.

7. Foucault, "Afterword," 222.

8. Arturo Escobar, "Beyond the Search for a Paradigm: Post-Development and Beyond," *Development* 43, no. 4 (2000), 14. Emphasis added by this author.

9. Glyn Williams, "Evaluating Participatory Development: Tyranny, Power and (Re)Politicisation," *Third World Quarterly* 25, no. 3 (2004): 557-78.

10. Andrea Cornwall and John Gaventa, *From Users and Choosers to Makers and Shapers: Repositioning Participation in Social Policy* (Brighton: IDS, 2001).

11. Barbara Cruikshank, *The Will to Empower: Democratic Citizens and Other Subjects* (Ithaca: Cornell University Press, 1999), 24.

12. Cruikshank, *The Will to Empower*, 1-9.

13. As pointed out by Giddens, social interactions entail "an interweaving of short-term purposes and long-term projects" (1995, 35), in contrast to the synchronic view prevailing among analysts to associate time with changes, thus underplaying recursive social practices across time. In the case of the donor-supported project, the long-standing struggles against *begaari* across generations constituted "storage capacity" (1995, 35) that *Tharus* drew upon to demand the "new rule" under the 2001 project. This leads us to conclude, by drawing on the two terms coined by Cowen and Shenton (1996), that it is crucial to pay attention to how "imminent" participatory development interventions engage with more "immanent" processes of day-to-day struggles.

14. In an interview held in May 2001, a senior manager in charge of this project insisted on the righteousness of the donor policy itself. At the same time, he admitted the

need occasionally to reach a compromise with those "politically motivated to disturb the smooth implementation of people's participation." The donor official was apparently wavering between his "technocratic" view equating development with a step-by-step sequence of taking decisions and executing them, and his admission of the inability to save his projects from being dragged into political bargaining.

15. Caroline M. Robb, *Can the Poor Influence Policy?: Participatory Poverty Assessments in the Developing World* (Washington, D.C.: The World Bank, 2002), xxvii.

16. Frances Cleaver, "The Social Embeddedness of Agency and Decision-making," in *Participation, from Tyranny to Transformation?: Exploring New Approaches to Participation in Development*, eds. Sam Hickey and Giles Mohan (London: Zed Books, 2004), 275.

17. Pierre Bourdieu, *The Logic of Practice* (London: Polity Press, 1990), 80-97.

18. Bourdieu, *The Logic of Practice*, 84.

# Appendix

## Policy Directive for "Participatory" River Control
### (Issued by the Minister of Water Resources in 1992)

9 November 1992

Note and Report

As far as the Jhapa district is concerned, [river control] works will be undertaken under the Chief of the Kankai Irrigation Commission [which supervises one project called the Kankai Irrigation Project] and the District Irrigation Office, according to the views expressed by the following committee.

Implementation Committee
1.  Chair of the District Development Committee
2.  Chief District Officer
3.  Local Development Officer
4.  One socially renowned and prestigious person
5.  Chief of the District Irrigation Office (member-secretary)
6.  The Member of Parliament of that area can be included for advice.

Local beneficiary groups will be actively involved in the implementation. The transportation of boulders is costly to control floodwater and riverbank erosion. Therefore, sandbags filled with sand will be used and held together by iron wire.

Laxman Prasad Ghimire
State Minister for Water Resources

# Bibliography

ADDCN. *Decentralization in Nepal: Prospects and Challenges (Findings and Recommendations of Joint HMG-Donor Review)*. Kathmandu: ADDCN (Association of District Development Committee of Nepal), 2001.

Alvesson, Mats, and Kaj Skoldberg. *Reflexive Methodology: New Vistas for Qualitative Research*. London: Sage, 2000.

Alvesson, Mats. *Postmodernism and Social Research*. Buckingham: Open University Press, 2002.

Apthorpe, Raymond, and Des Gasper, eds. *Arguing Development Policy: Frames and Discourses*. London: Frank Cass, 1996.

Arce, Alberto. "Experiencing the Modern World: Individuality, Planning and the State." Pp. 103-17 in *Resonances and Dissonances in Development: Actors, Networks and Cultural Reportoires*, edited by Paul Hebinck and Gerald Verschoor. Assen: Royal Van Gorcum, 2001.

———. "Re-Approaching Social Development: A Field of Action between Social Life and Policy Process." *Journal of International Development* 15 (2003): 845-61.

Arce, Alberto, Magdalena Villarreal, and Pieter De Vries. "The Social Construction of Rural Development: Discourses, Practices and Power." Pp. 152-71 in *Rethinking Social Development: Theory, Research and Practice*, edited by David Booth. Essex: Longman, 1994.

Bachrach, Peter, and Morton S. Baratz. *Power and Poverty: Theory and Practice*. New York: Oxford University Press, 1970.

Barker, Philip. *Michel Foucault: An Introduction*. Edinburgh: Edinburgh University Press, 1998.

Beetham, David. *The Legitimation of Power*. London: Macmillan, 1991.

*Bhumishdhar Bibhag. Bardiya Jillama Bhmishdhar*. Kathmandu: *Sree Panchko Sarkarko Chhapakhana* (Department of Land Reform. *Land Reform in Bardiya District*. Kathmandu: His Majesty's Government), 1966

Blaikie, Piers, Terry Cannon, Ian Davis, and Ben Wisner. *At Risk: Natural Hazards, People's Vulnerability, and Disasters*. London: Routledge, 1994.

Borre, Ole, Sushil R. Panday, and Chitra K. Tiwari. *Nepalese Political Behaviour*. New Delhi: Sterling Publishers, 1994.

Bourdieu, Pierre. *Outline of a Theory of Pr* 133 Cambridge: Cambridge University Press, 1977.

———. *The Logic of Practice*. London: Polity Press, 1990.

Clay, Edward J., and Bernard B. Schaffer. *Room for Manoeuvre: An Exploration of Public Policy in Agricultural and Rural Development*. London: Heinemann Educational Books, 1984.

Cleaver, Frances. "The Social Embeddedness of Agency and Decision-making." Pp. 271-78 in *Participation, from Tyranny to Transformation?: Exploring New Approaches to Participation in Development*, edited by Sam Hickey and Giles Mohan. London: Zed Books, 2004.

Clegg, Stewart R. *Frameworks of Power*. London: Sage Publications, 1989.

Cooke, Bill, and Uma Kothari. "The Case for Participation as Tyranny." Pp. 1-15 in *Participation: The New Tyranny?*, edited by Bill Cooke and Uma Kothari. London: Zed Books, 2001.

Cornwall, Andrea. *Beneficiary, Consumer, Citizen: Perspectives on Participation for Poverty Reduction*. Stockholm: SIDA, 2000.

———. *Making Spaces, Changing Places: Situating Participation in Development* (IDS Working Paper 170). Brighton: IDS, 2002.

Cornwall, Andrea, and John Gaventa. *From Users and Choosers to Makers and Shapers: Repositioning Participation in Social Policy* (IDS Working Paper 127). Brighton: IDS, 2001.

Cowen, Michael, and Robert W. Shenton. *Doctrine of Development*. London: Routledge, 1996.

CPN-UML. *People's Multi-Party Democracy: Programme of Nepalese Revolution*. Kathmandu: CPN-UML (Communist Party of Nepal - United Marxist-Leninist), 1998.

Crewe, Emma, and Elizabeth Harrison. *Whose Development? An Ethnography of Aid*. London: Zed Books, 1998.

Cruikshank, Barbara. *The Will to Empower: Democratic Citizens and Other Subjects*. Ithaca: Cornell University Press, 1999.

Dahal, Dev R. *The Challenge of Good Governance: Decentralization and Development in Nepal*. Kathmandu: Centre for Governance and Development Studies, 1996.

Dahl, Robert A. *Who Governs?: Democracy and Power in an American City*. New Haven: Yale University Press, 1961.

Dakhal, Suresh, Janak Rai, Dambar Chemchong, Dhruba Maharajan, Pranita Pradhan, Jagat Maharajan, and Shreeram Chaudhary. *Issues and Experiences: Kamaiya System, Kanara Andolan and Tharus in Bardiya*. Kathmandu: SPACE (Society for Participatory Cultural Education), 2000.

Davies, Charlotte A. *Reflexive Ethnography: A Guide to Researching Selves and Others*. London: Routledge, 1999.

Digester, Peter. "The Fourth Face of Power." *The Journal of Politics* 54, no. 4 (1992): 977-1007.

Dreyfus, Hubert L., and Paul Rabinow. *Michel Foucault: Beyond Structuralism and Hermeneutics*. Brighton: The Harvester Press, 1982.

Edelman, Murray. *Political Language: Words That Succeed and Policies That Fail*. New York: Academic Press, 1977.

Escobar, Arturo. *Encountering Development: The Making and Unmaking of the Third World*. Princeton: Princeton University Press, 1995.

———. "Beyond the Search for a Paradigm: Post-Development and Beyond." *Development* 43, no. 4 (2000): 11-14.

Fagan, Honor. "Cultural Politics and (Post) Development Paradigm(S)." Pp. 178-95 in *Critical Development Theory: Contributions to a New Paradigm*, edited by Robaldo Munck and Denis O'Hearn. London: Zed Books, 1999.

Fairclough, Norman. *Discourse and Social Change*. Cambridge: Polity Press, 1992.

Ferguson, James. *The Anti-Politics Machine: "Development," Depoliticization, and Bureaucratic Power in Lesotho*. Cambridge: Cambridge University Press, 1990.

Foucault, Michel. *Discipline and Punish: The Birth of the Prison*. London: Penguin Books, 1976.

———. *Lecture*. Pasquino: College de France, 1978.

———. *The History of Sexuality, Vol.1: Introduction*. Harmondsworth: Penguin, 1979.

———. "Two Lectures." Pp. 78-108 in *Power/Knowledge: Selected Interviews and Other Writings 1972-1977 Michel Foucault*, edited by Colin Gordon. New York: Harvester Wheatsheaf, 1980.

———. "Afterword: The Subject and Power." Pp. 208-26 in *Michel Foucault: Beyond Structuralism and Hermeneutics*, edited by Hubert L. Dreyfus and Paul Rabinow. Brighton: The Harvester Press, 1982.

———. *The Archaeology of Knowledge*. London: Tavistock, 1986.

————. "Questions of Methods." Pp. 73-86 in *The Foucault Effect: Studies in Governmentality*, edited by Graham Burchell, Colin Gordon and Peter Miller. Chicago: The University of Chicago Press, 1991a.

————. "Governmentality." Pp. 87-104 in *The Foucault Effect: Studies in Governmentality*, edited by Graham Burchell, Colin Gordon and Peter Miller. Chicago: The University of Chicago Press, 1991b.

Gardner, Katy, and David Lewis. *Anthropology, Development and the Post-Modern Challenge*. London: Pluto Press, 1996.

Gasper, Des. "Essentialism in and About Development Discourse." Pp. 149-76 in *Arguing Development Policy: Frames and Discourses*, edited by Raymond Apthorpe and Des Gasper. London: Frank Cass, 1996.

Gaventa, John, and Camilo Valderrama. *Participation, Citizenship and Local Governance: Background Paper for Workshop, Strengthening Participation in Local Governance*. Brighton: IDS, 1999.

Gaventa, John. "Introduction: Exploring Citizenship, Participation and Accountability." *IDS Bulletin* 33, no. 2 (2002): 1-11.

Giddens, Anthony. *Central Problems in Social Theory: Action, Structure and Contradiction in Social Theory*. London: Macmillan, 1979.

————. *Profiles and Critiques in Social Theory*. London: Macmillan, 1982.

————. *Constitution of Society*. Cambridge: Policy Press, 1984.

Gordon, Colin. "Governmental Rationality: An Introduction." Pp. 1-51 in *The Foucault Effect: Studies in Governmentality*, edited by Graham Burchell, Colin Gordon and Peter Miller. Chicago: The University of Chicago Press, 1991.

Grillo, Ralph D. "Discourse of Development: The View from Anthropology." Pp. 1-34 in *Discourse of Development: Anthropological Perspectives*, edited by Ralph D. Grillo and Roderick. L. Stirrat. Oxford: Berg, 1997.

Grindle, Merilee S., and John W. Thomas. *Public Choices and Policy Change: The Political Economy of Reform in Developing Countries*. Baltimore: Johns Hopkins University Press, 1991.

Guneratne, Arjun. "Modernization, the State, and the Construction of a Tharu Identity in Nepal." *The Journal of Asian Studies* 57, no. 3 (1998): 749-73.

Hachhethu, Krishna. "Nepali Politics: Political Parties, Political Crisis and Problems of Governance." Pp. 90-116 in *Domestic Conflict and Crisis of Governability in Nepal*, edited by Dhruba Kumar. Kathmandu: CNAS (Centre for Nepal and Asian Studies), 2000.

Hajer, Maarten. A. *The Politics of Environmental Discourse: Ecological Modernization and the Policy Process*. Oxford: Clarendon Press, 1995.

Hammersley, Martyn, and Paul Atkins. *Ethnography: Principles in Practice*. London: Routledge, 1995.

Hayward, Clarissa R. "De-Facing Power." *Polity* 31, no. 1 (1998): 1-22.

Hebinck, Paul, Janden Ouden, and Gerald Verschoor. "Past, Present, and Future: Long's Actor-Oriented Aproach at the Interface." Pp. 1-16 in *Resonances and Dissonances in Development: Actors, Networks and Cultural Reportoires*, edited by Paul Hebinck and Gerald Verschoor. Assen: Royal Van Gorcum, 2001.

Hickey, Sam, and Giles Mohan. "Towards Participation as Transformation: Critical Themes and Challenges." Pp. 3-24 in *Participation, from Tyranny to Transformation?: Exploring New Approaches to Participation in Development*, edited by Sam Hickey and Giles Mohan. London: Zed Books, 2004a.

———, eds. *Participation, from Tyranny to Transformation: New Approaches to Participation in Development*. London: Zed Books, 2004b.

Hill, Michael. *The Policy Process in the Modern State*. 3d ed. Hertfordshire: Prentice Hall/Harvester Wheatsheaf, 1997.

Howarth, David. *Discourse*. Buckingham: Open University Press, 2000.

Howarth, David, and Yannis Stavrakakis. "Introducing Discourse Theory and Political Analysis." Pp. 1-34 in *Discourse Theory and Political Analysis: Identities, Hegemonies and Social Change*, edited by David Howarth, Aletta J. Norval and Yannis Stavrakakis. Manchester: Manchester University Press, 2000.

Hoy, David C. "Power, Repression, Progress: Foucault, Lukes, and the Frankfurt School." Pp. 123-47 in *Foucault: A Critical Reader*, edited by David C. Hoy. Oxford: Basil Blackwell, 1986.

Kapoor, Ilan. "The Devil's in the Theory: A Critical Assessment of Robert Chambers' Work on Participatory Development." *Third World Quarterly* 23, no. 1 (2002): 101-17.

Kiely, Ray. "The Last Refuge of the Noble Savage?: A Critical Assessment of Post-Development Theory." *The European Journal of Development Research* 11, no. 1 (1999): 30-55.

Laclau, Ernesto. *New Reflections on the Revolution of Our Time*. London: Verso, 1990.

Laclau, Ernesto, and Chantal Mouffe. *Hegemony and Socialist Strategy*. 2d ed. London: Verso, 2001.

Lam, Wai F. *Governing Irrigation Systems in Nepal: Institutions, Infrastructure, and Collective Action.* Oakland: ICS Press, 1998.

Li, Tania M. "Images of Community: Discourse and Property Relations." *Development and Change* 27, no. 3 (1996): 501-27.

Lipsky, Michael. *Street-Level Bureaucracy: Dilemmas of the Individual in Public Services.* New York: Russell Sage Foundation, 1980.

Long, Norman. "From Paradigm Lost to Paradigm Regained." Pp. 16-43 in *Battlefields of Knowledge: The Interlocking of Theory and Practice in Social Research and Development,* edited by Norman Long and Ann Long. London: Routledge, 1992.

———. "Exploring Local/Global Tranformations: A View from Anthropology." Pp. 184-201 in *Anthropology, Development and Modernities: Exploring Discourses, Counter-Tendencies and Violence,* edited by Alberto Arce and Norman Long. London: Routledge, 2000.

———. *Development Sociology: Actor Perspectives.* London: Routledge, 2001.

Lukes, Steven. *Power: A Radical View.* London: Macmillan, 1974.

———. *Power: A Radical View.* 2d ed. Basingstoke: Palgrave Macmillan, 2005.

Martinussen, John. *Local Authorities in Nepal: An Assessment of Their Present Position and Proposal for Strengthening of Democracy at the Local Level.* Kathmandu: Local Development Training Academy, 1993.

Masaki, Katsuhiko. "The 'Transformative' Unfolding of 'Tyrannical' Participation: The Corvée Tradition and Ongoing Local Politics in Western Nepal." Pp.125-39 in *Participation, from Tyranny to Transformation?: Exploring New Approaches to Participation in Development,* edited by Sam Hickey and Giles Mohan. London: Zed Books, 2004.

———. "The Oppression/Emancipation Nexus in Ongoing Power Struggles: Village-Power Dynamics in Western Nepal." *Journal of Development Studies* 42, no. 5 (2006): 721-38.

Moore, Donald S. "The Crucible of Cultural Politics: Reworking 'Development' in Zimbabwe's Eastern Highlands." *American Ethnologist* 26, no. 3 (2000): 654-89.

Mosse, David. "The Ideology and Politics of Community Participation: Tank Irrigation Development in Colonial and Contemporary Tamil Nadu." Pp. 255-92 in *Discourse of Development : Anthropological Perspective,* edited by Ralph D. Grillo and Roderick. L. Stirrat. Oxford: Berg, 1997.

————. ""People's Knowledge", Participation and Patronage: Operations and Representation in Rural Development." Pp. 16-35 in *Participation: The New Tyranny*, edited by Bill Cooke and Uma Kothari. London: Zed Books, 2001.

Nederveen Pieterse, Jan. *Development Theory: Deconstructions/Reconstructions*. London:
Sage, 2001.

Pandey, Devendra R. *Nepal's Failed Development: Reflections on the Mission and the
Maladies*. Kathmandu: Nepal South Asia Centre, 1999.

Parsons, Wayne. *Public Policy: An Introduction to the Theory and Practice of Policy
Analysis*. Aldershot: Edward Elgar, 1995.

Pigg, Stacy L. "Investing Social Categories through Place: Social Representations and
Development in Nepal." *Comparative Study of Society and History* 34, no. 3
(1992): 491-513.

————. "Unintended Consequences: The Ideological Impact of Development in Nepal."
*South Asia Bulletin* 13, no. 1&2 (1993): 45-58.

Potter, Jonathan, and Margaret Wetherell. *Discourse and Social Psychology: Beyond Attitudes and Behaviour*. London: Sage, 1987.

Poudyal Chhetri, Meen B., and Damodar Bhattarai. *Mitigation and Management of Floods
in Nepal*. Kathmandu: Format Printing Press, 2001.

Robb, Caroline M. *Can the Poor Influence Policy?: Participatory Poverty Assessments in
the Developing World*. Washington, D.C.: The World Bank, 2002.

Sabatier, Paul A., and Hank C. Jenkins-Smith. *Policy Change and Learning: An Advocacy
Coalition Approach*. Boulder: Westview Press, 1993.

Schaffer, Bernard B. "Towards Responsibility: Public Policy in Concept and Practice." Pp.
142-90 in *Room for Manoeuvre: An Exploration of Public Policy in Agricultural
and Rural Development*, edited by Edward J. Clay and Bernard B. Schaffer.
London: Heinemann Educational Books, 1984.

Scheurich, James J. *Research Method in the Postmodern*. London: Routledge/Falmer,
1997.

Schon, Donald A., and Martin Rein. *Frame Reflection: Toward the Resolution of Intrac-
trable Policy Controversies*. New York: Basic Books, 1994.

Scott, James C. *Domination and the Arts of Resistance: Hidden Transcripts*. New Haven:
Yale University Press, 1990.

Sharp, Joanne P., Paul Routledge, Chris Philo, and Ronan Paddison. "Entanglements of
Power: Geographies of Domination/Resistance." Pp. 1-42 in *Entanglements of*

*Power: Geographies of Domination/Resistance*, edited by Joanne P. Sharp, Routledge Paul, Chris Philo and Ronan Paddison. London: Routledge, 2000.

Shore, Cris, and Susan Wright, eds. *Anthropology of Policy: Critical Perspectives on Governance and Power*. London: Routledge, 1997.

Shrestha, Tulsi N. *The Implementation of Decentralisation Scheme in Nepal: An Assessment and Lessons for Future*. Kathmandu: Joshi Publications, 1999.

Smith, Martin J. *Pressure, Power and Policy: State Autonomy and Policy Neworks in Briton and the United States*. New York: Harvester Wheatsheaf, 1993.

Steier, Frederick. "Introduction: Research as Self-Reflexivity, Self-Reflexivity as Social Process." Pp. 1-11 in *Research and Reflexivity*, edited by Frederick Steier. London: Sage, 1991a.

———. "Reflexivity and Methodology: An Ecological Constructionism." Pp. 163-85 in *Research and Reflexivity*, edited by Frederick Steier. London: Sage, 1991b.

Stirrat, Roderick L. "The New Orthodoxy and Old Truths: Participation, Empowerment and Other Buzz Words." Pp. 67-92 in *Assessing Participation: A Debate from South Asia*, edited by Sunil Bastian and Nicola Bastian. Delhi: Konark Publishers, 1996.

Storey, Andy. "Post-Development Theory: Romanticism and Pontius Pilate Politics." *Development* 43, no. 4 (2000): 40-46.

Tendler, Judith. *Good Government in the Tropics*. Baltimore: The Johns Hopkins University Press, 1997.

Williams, Glyn. "Evaluating Participatory Development: Tyranny, Power and (Re)Politicisation." *Third World Quarterly* 25, no. 3 (2004): 557-78.

Wood, Geof, ed. *Labelling in Development Policy*. London: Sage, 1985.

# Index

actor-oriented approach, 24-26, 29, 30

agency: agency/structure nexus, 21-23, 38n29; citizen participation and, 30-32, 123-25; disciplinary power and, 20-21, 23; in policymaking, 24-25

aid agency. *See* donor agency

antagonisms of strategies, 20, 108, 117. *See also* social antagonism

anthropology of policy, 3, 6, 26-29, 40n67, 124; and river control policy, 48-50, 122-23

Arce, Alberto, 25-26

ascending/descending analysis, 20, 37n15; of policymaking, 6, 12, 13, 128; of river control policy, 57-58, 73-74, 91-92, 110-11, 121-22; at the village level, 87, 106-7

Bachrach and Baratz, 18

Bardiya district, 6-7, 12, 62; fieldwork in 7-11; land reform in, 64, 75n8, 78-79, 96-97; political history of, 63-64

*begaari. See* corvée tradition

Bourdieu, Pierre, 10, 21, 38n29, 85, 116-17, 128

caste, 63, 74n5, 78, 92n2, 92n4, 96. *See also dalit*

citizen participation, 2, 3, 13, 29-32, 123-125, 126, 127; technologies of

citizenship, 31, 124; citizen-subject, 31, 124

Clay and Schaffer, 25

Cleaver, Frances, 128

Clegg, Stewart R., 25

Cornwall and Gaventa 124

Cornwall, Andrea, 2

corvée tradition, 9-11, 64-66, 75n9, 75n11, 126; historical origin of, 63; precariousness of, 66-69, 83-90, 101-9, 117-18, 125-27; and river control policy, 12, 61

CPN-UML (Communist Party of Nepal-Unified Marxist-Leninist), 50, 51, 52, 59n12, 59n13, 59n14, 60n28, 63, 73, 75n12, 80, 87, 89, 91, 92n7, 93n10, 97, 101, 106, 111n4, 111n5, 111n6, 112n9

Cruikshank, Barbara, 31, 124

cultural politics, 3, 28-29, 31, 124, 127-29; and river control policy, 6, 49-50, 91, 110, 122-23

Dahl, Robert A., 18

*dalit*, 101, 102-3, 116

D&P (decentralization and participation) policy, 12, 45-47, 48; and corvée tradition, 65, 66, 75n11, 117; and Nepal's policymakers, 50-52, 53-54, 59n13, 59n15, 67, 70-71

# About the Author

**Katsuhiko Masaki** is Associate Professor at the Department of Studies on Global Citizenship of Seisen University in Tokyo, Japan. He holds a Ph.D. from the Institute of Development Studies (IDS) at the University of Sussex, Brighton, UK. Prior to becoming an academic in 2003, Professor Masaki worked for several international development agencies, including the United Nations Development Programme (UNDP). While teaching at Seisen University, he not only continues to undertake consultancy assignments with various agencies, but also contributes articles to textbooks for practioners, including a chapter in a reading packet for social development advisers of the UK's Department of International Development (DFID). He has published widely, both inside and outside Japan, on issues of power, participation, and development policy. His major publications include: "The Oppression/Emancipation Nexus in Ongoing Power Struggles," in the *Journal of Development Studies* 42, no. 5 (2006), and "The 'Transformative' Unfolding of 'Tyrannical' Participation," in *Participation: From Tyranny to Transformation?*, eds. Sam Hickey and Giles Mohan (2004).